The Influence of Environment on Mechanical Characteristics of Key Semiconductor Components

Joshua Pelleg

The Influence of Environment on Mechanical Characteristics of Key Semiconductor Components

Springer

Joshua Pelleg
Department of Materials Engineering
Ben Gurion University of the Negev
Be'er Sheva, Israel

ISBN 978-3-031-34529-6 ISBN 978-3-031-34530-2 (eBook)
https://doi.org/10.1007/978-3-031-34530-2

This Springer imprint is published by the registered company Springer Nature Switzerland AG
The registered company address is: Gewerbestrasse 11, 6330 Cham, Switzerland

In memoriam-Ada Pelleg

03/09/1934–27/2/1022

While launching this book Ada my wife, blessed be her memory, suddenly passed away. This book is dedicated to her. Her inspiration in initiating this book and writing this and my other books is greatly missed.

The fragility of life is magnified on personal loss.

Joshua Pelleg

To my late wife Ada and my children, Dina, and her late husband Gidon Barak, Ruth and Christer Kallevag, Shlomit and Asher Pelleg and to my grandchildren: Roy, Tal, Rotem and Noa Barak; Ella and Maya Kallevag; and Ofir and Ori Pelleg.

Preface

Chapter 1 sets the framework of the environmental constituents considered in this book and their deleterious effect on material properties, in particular those of mechanical. Often these are hostile not only to the tested material, but also to the health of persons who handle the material. The environmental constituents are listed in this chapter which includes oxygen, nitrogen hydrogen, air, moisture in air and vacuum. Some of these have proven to directly affect climate due to greenhouse gasses emission and it is urgently required that economy asses their global environmental impacts, find and suggest alternate use for their production, in order to preserve life on earth. The subject of this book is not "life on earth", but rather of the likely harmful (or beneficial) effect of these gases on the mechanical properties of material as outlined in the content. Chapter 2 considers elastic properties with yielding as the demarcation point between elasticity and plasticity, while Chap. 3 deals with plastic deformation discussing tensile and compressive test in the respective sections. Defects such as dislocations that make plastic deformation and fracture occur at lower stresses than the theoretical are considered in Chap. 4, which also includes a section on point defects. Defects are essential features of every material, and semiconductors are no exception. Point defects and dislocations are the defects responsible for the behavior of a specific material, and in this section the effect of environment on this behavior is discussed. Chapters 5 and 6 are related to hardness and brittle behavior of semiconductors as effected by various environments. The hardness test is used instead of the tensile test as it is simpler, cheaper, and quicker to perform. As there is a direct correlation between hardness and tensile strength, it is used to evaluate tensile strength. Recall that brittleness describes the property of a material that fractures having a little tendency to deform before failure when subjected to stress. Generally, brittle materials are characterized by little deformation, poor resistance to high compressive strength, and low tensile strength. Impact testing covered in Chap. 7 illustrates by the contraction at the notch of a broken specimen the deformation tendency whether brittle or ductile failure. Deformation is directly associated with dislocations, but an additional deformation process is described, namely, by twinning which is considered in Chap. 8. Chapter 9 is devoted to time dependent deformation, namely creep, which is augmented by temperature.

The statically loaded material, say by simple tension, elongates with the elapse of time, until fracture sets in and the process occurs faster above room temperature and increases with increased temperature. To understand fatigue is in particular important, since about 90% of all failures occur by fatigue and thus critical to the fatigue life of the semiconducting constituents also. The origin of the damage by fatigue may start from a micro-crack. Fatigue behavior also known as cyclic deformation is discussed in Chap. 10. Fracture sets in material either by not being able to carry the applied load or by stress rupture and is considered in Chap. 11. Fracture may be brittle or ductile. Mechanical behavior of small size semiconductor specimens where the dimensions are in the nano scale is discussed in Chap. 12. Deformation of nanostructures is expected to be basically different than the macro-size counterpart ones, and it might be expected for improved mechanical properties.

All the above chapters will be discussed in the light of the influence of the surrounding media indicated in the title of this book.

In the closing, I would like to express my gratitude to all publishers and authors for permission to use and reproduce some of their illustrations and microstructures.

Finally, without the tireless devotion, help, understanding, and unlimited patience of my wife Ada, who recently passed away, I could never have started and completed this book, despite my decades of teaching in this field; her encouragement was essential, and her helpful attitude was instrumental in inspiring to write this book.

Be'er Sheva, Israel Joshua Pelleg

Contents

1 **The Environmental Challenge in Semiconductor Production** 1
1.1 Effect of Oxygen ... 2
1.1.1 In Si ... 2
1.1.2 In SiC ... 5
1.2 Effect of Moisture ... 6
1.2.1 In Si ... 6
1.2.2 In SiC ... 7
1.2.3 In GaAs ... 13
References ... 16

2 **Elastic Properties** ... 17
2.1 The Effect of Oxygen ... 17
2.1.1 In SiC ... 17
Reference ... 23

3 **Plastic Deformation** ... 25
3.1 The Effect of Oxygen ... 25
3.1.1 In Si ... 25
3.1.2 In SiC ... 28
3.2 The Effect of Moisture ... 31
3.2.1 In Si ... 31
3.2.2 In SiC ... 34
3.3 Tensile Deformation, The Effect of Environment ... 42
3.3.1 The Effect of Oxygen ... 42
3.3.2 The Effect of Moisture ... 47
3.4 Compressive Deformation the Effect of Environment ... 50
3.4.1 The Effect of Oxygen ... 50
References ... 61

4 Hardness and Tensile Properties. The Effect of Environment 63
4.1 The Effect of Oxygen ... 63
4.1.1 In Ge0.5Si0.5 ... 63
4.2 The Effect of Moisture ... 67
4.2.1 In Si ... 67
References ... 69

5 Strength—The Effect of Environment ... 71
5.1 The Effect of Oxygen ... 71
5.1.1 In Si ... 71
5.1.2 In SiC ... 73
5.2 The Effect of Moisture ... 76
5.2.1 In SiC ... 76
5.2.2 In GaAs ... 78
References ... 81

6 Brittle Behavior-The Effect of Environment ... 83
6.1 The Effect of Oxygen ... 83
6.1.1 In SiC ... 83
6.1.2 In GaAs ... 85
6.2 The Effect of Moisture ... 90
6.2.1 In SiC ... 90
6.2.2 In GaAs ... 94
References ... 104

7 Creep–Time Dependent Deformation. Effect of Environment 105
7.1 The Effect of Oxygen ... 105
7.1.1 In SiC ... 105
References ... 108

8 Fatigue-Cyclic Deformation. Effect of Environment ... 109
8.1 The Effect of Oxygen ... 109
8.1.1 In SiC ... 109
8.2 The Effect of Moisture ... 115
8.2.1 In SiC ... 115
References ... 120

9 Fracture—Effect of Environment ... 121
9.1 The Effect of Oxygen ... 121
9.1.1 In Si ... 121
9.1.2 In SiC ... 126
9.1.3 In GaAs ... 129
References ... 131

10 Twinning–Effect of Environment ... 133
10.1 Introduction ... 133
10.2 The Effect of Oxygen ... 135
10.2.1 In Si ... 135
10.2.2 In SiC ... 137
10.2.3 In GaAs ... 138
References ... 147

11 Fracture Toughness–Effect of Environment ... 149
11.1 Introduction ... 149
11.2 Impact Test ... 151
11.2.1 Notch Effect ... 152
11.3 The Effect of Oxygen ... 154
11.3.1 In Si ... 154
References ... 155

12 Deformation in Nanostructure–Effect of Environment ... 157
12.1 Introduction ... 157
12.2 Nano-Wear Deformation ... 158
12.2.1 Effect of Oxygen ... 158
12.3 Tensile Deformation in Nanostructure–Effect of Environment ... 163
12.3.1 Effect of Oxygen ... 163
12.3.2 Effect of Moisture ... 175
12.4 Creep in Nanomaterials Effect of Environment ... 177
12.4.1 Effect of Oxygen ... 177
12.4.2 Effect of Moisture ... 179
12.5 Fatigue in Nanomaterials-Effect of Environment ... 185
12.5.1 Effect of Oxygen ... 185
12.5.2 Effect of Moisture ... 190
12.6 Fracture in Nanomaterials-Effect of Environment ... 198
12.6.1 Effect of Oxygen ... 198
12.6.2 Effect of Moisture ... 209
References ... 215

Index ... 217

About the Author

Joshua Pelleg received his B.Sc. in Chemical Engineering at the Technion–Israel Institute of Technology, Haifa, Israel, M.Sc. in Metallurgy at the Illinois Institute of Technology, Chicago, IL, and Ph.D. in Metallurgy at the University of Wisconsin, Madison, WI. He has been in the Ben-Gurion University of the Negev (BGU) Materials Engineering Department in Beer-Sheva, Israel since 1970, and was among the founder of the department, and served as its second chairman. Professor Pelleg was the recipient of the Samuel Ayrton Chair in Metallurgy. He specializes in the mechanical properties of materials and the diffusion and defects in solids. He has chaired several university committees and served four terms as the Chairman of Advanced Studies at Ben-Gurion University of the Negev. Prior to his work at BGU, Pelleg acted as Assistant Professor and then Associate Professor in the Department of Materials and Metallurgy at the University of Kansas, Lawrence, KS. Professor Pelleg was also a Visiting Professor: in the Department of Metallurgy at Iowa State University at the Institute for Atomic Research, US Atomic Energy Commission, Ames, IA; at McGill University, Montreal, QC; at the Tokyo Institute of Technology, Applied Electronics Department, Yokohama, Japan; and in Curtin University, Department of Physics, Perth, Australia. His nonacademic research and industrial experience includes: Chief Metallurgist in Urdan Metallurgical Works Ltd., Netanya, Israel; Research Engineer in International Harvester Manufacturing Research, Chicago, IL; Associate Research Officer for the National

Research Council of Canada, Structures and Materials, National Aeronautical Establishment, Ottawa, ON; Physics Senior Research Scientist, Nuclear Research Center, Beer-Sheva, Israel; Materials Science Division, Argonne National Labs, Argonne, IL; Atomic Energy of Canada, Chalk River, ON; Visiting Scientist, CSIR, National Accelerator Centre, Van de Graaf Group Faure, South Africa; Bell Laboratories, Murray Hill, NJ; and GTE Laboratories, Waltham, MA. His current research interests are mechanical properties, diffusion in solids, thin film deposition and properties (mostly by sputtering) and the characterization of films, among them various silicides. His research activity is summarized in his publications in various scientific journals and in 11 books published by Springer and Elsevier.

Chapter 1
The Environmental Challenge in Semiconductor Production

In the following I will consider the surrounding media as it appears in the title of this book as environment, since some of them (excluding clearly vacuum) appear among the constituents of the greenhouse effect gases directly and some are involved indirectly in greenhouse gas formation associated with global climate change. Vacuum is considered among the mentioned gases for comparison purposes for obtaining the properties without the influence of these atmospheric gases.

Among all industries producing various components or finished products, the transistor industry -despite the relatively small size products-stands out as a leading manufacturing enterprise. It is however the most polluting industry, and therefore their impact on the environment in making computer chip and other devices has to be reduced. Moreover, in producing millions of chips yearly by a typical semiconductor factory, gulping huge amounts of water is involved and also large quantities of many toxic chemicals all of which can harm the environment are also used. All these including the rinsing water in each step of chip making is disposed as waste comprising the pollution. Continued emission of various perfluorinated compounds (PFCs), a group of global warming gases, which is used in the production (etching) of semiconductors have high global warming potentials (GWP) and extremely long atmospheric lifetimes contributes to the greenhouse effect by the known gaseous compounds. It is essential to remove all these factors that contribute to the greenhouse effect and the global warning, not only by alternative production methods and the use -if at all possible-more friendly, less toxic chemicals but also full recycling of water (avoiding, river, stream, and other water resources contamination). It is a big challenge to perform the necessary steps for achieving a positive attitude from the manufacturers by reducing environmental impact on semiconductor production, since the required changes add significantly to the manufacturing cost. It is paradoxical that all the good things that men receive from the semiconductor industry is accompanied with all the bad consequences and impacts on global warming, human and animal health and environmental pollution.

J. Pelleg, *The Influence of Environment on Mechanical Characteristics of Key Semiconductor Components*, https://doi.org/10.1007/978-3-031-34530-2_1

The environmental constituents listed above, namely, oxygen, nitrogen, hydrogen, air, moisture in air and vacuum are not directly related to global warming, but they constitute the composition, -sometimes in very small quantities-, of the atmosphere. Their effect, -harmful or beneficial-, on the material properties operating in such or similar environment (atmosphere) is of interest in order to assess the probable outcome of their influence on the strength properties and lifetime. The application of some of these atmospheric constituents as supporting media of their effect on strength properties will be considered, depending on their availability in the published literature. One should be aware, that only some of these constituents such as moisture are really described in publications concerning their effect. Thus, the tendency of this communication is limited to the available data, and not necessarily will be covering all or partially specific properties indicated in the content. Specific mechanical properties are discussed in the appropriate chapters, but here general behavior of these constituents is discussed. The materials affected by the indicated gasses used here are Si, Ge, SiC and GaAs.

1.1 Effect of Oxygen

1.1.1 In Si

Low oxygen content floating zone (FZ) silicon wafer used for insulated gate bipolar transistors (IGBTs) was tested at high temperature by bending for the evaluation of the critical stress (temperature dependent) and the oxygen concentration distribution along the depth from the wafer surface. It has been found that the critical stress does not depend on temperature above 973 K but, depends significantly on the oxygen concentration distribution along the depth direction from the wafer surface, Furthermore, it was observed that controlling the oxygen concentration below the wafer surface by thermal treatment suppressed slip generation during the device process, which can contribute to solving the problem of wafer bending and current leakage. The slip generation depth (from 3.3 to 7.3 mm) is shown in Table 1.1. The dislocation locking stress is high and thus slip did not occur because the high oxygen concentration below the surface of the wafer was high. Oxygen concentration C_0 was calculated by Eq. (1.1) given below as

(Eq. (6) in Table 1.1 is in this paper is Eq. 1.1)

$$\tau_u = 5.3x10^{-5} C_0 \exp\left(\frac{0.88eV}{kT}\right) \tag{1.1}$$

Basically, this equation is

$$\tau_u = AC_{0\,\exp}\left(-\frac{E_a}{kT}\right) \tag{1.2}$$

Table 1.1 Summary of critical shear stress, oxygen concentration and depth of slip generation at each pre-annealing condition. J. Fujise, B. Ko, T. Ono and M. Tanaka, ECS Journal of Solid-State Science and Technology, 9, 104,002 (2020). Open access

Pre-annealing temperature (K)	Pre-annealing time	Critical shear stress (MPa)	Calculated Co by Eq. 6	Depth of slip generation
1373	30 min	2.5	1.7 × 10^{17} atoms cm"3	3.3/jm
1373	2 h	2.9	2.0 × 10^{17} atoms cm"3	5.0 fum
1373	5 h	3.7	2.5 × 10" atoms cm"3	6.1 ^m
1423	30 min	2.8	1.9 × 10^{17} atoms cm"3	5.0 ^m
1423	2 h	4.2	2.9 × 10^{17} atoms cm"3	7.3 ^m
1423	5 h	5.9	4.0 × 10^{17} atoms cm"1	6.7 /j m

where A is a temperature independent factor, $C_0 = 1 \times 10^{17}$ atoms cm^{-3} is the oxygen concentration and $E_a$is the activation energy for releasing a dislocation from locking oxygen atoms.

The values of A and E_a were obtained by fitting τ_u curves to the experimental data, then Eq. 1.1 was obtained.

The oxygen concentration was measured by SIMS. The depth profiles of oxygen concentration from wafer surface region after pre-annealing is shown in Fig. 1.1. The

depth penetration of oxygen concentration is higher in the specimens at the higher temperature which indicates that higher pre-annealing temperature increases oxygen concentration near the surface. It also indicates that the longer the pre-annealing time the higher the oxygen concentration in the deep region. The average depth of slip generation for a 0.98 N indent is 5.6 μm. Figure 1.2 shows the relationship between the critical shear stress and the oxygen concentration at a depth of 5.6 μm in each indent. TEM image just under the indent of 0.98 N is shown in Fig. 1.3. Since the loads of all indentation were the same, i.e., 0.98 N the depth of slip generation should also be the same and 3.3 to 7.3 μm is the variation in the depth of slip generation with an average of 5.6 mm.

It is expected that the failure during the device process can be reduced by paying attention to the oxygen concentration distribution in the depth direction. For example, in a low-oxygen concentration wafer, it is possible to prevent slip generation during the device process by performing pre-annealing over a long time at a high temperature to increase the oxygen concentration below the wafer surface. Increasing oxygen content

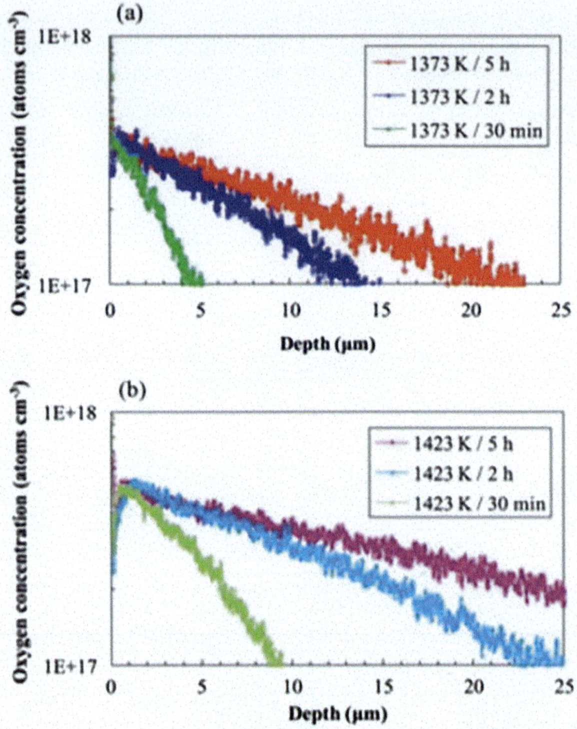

Fig. 1.1 Depth profiles of oxygen concentration from wafer surface region after each pre-annealing condition described in Table 1.1 **a** 1373 K, and **b** 1423 K. J. Fujise, B. Ko, T. Ono and M. Tanaka, ECS Journal of Solid-State Science and Technology,9, 104,002 (2020). Open access

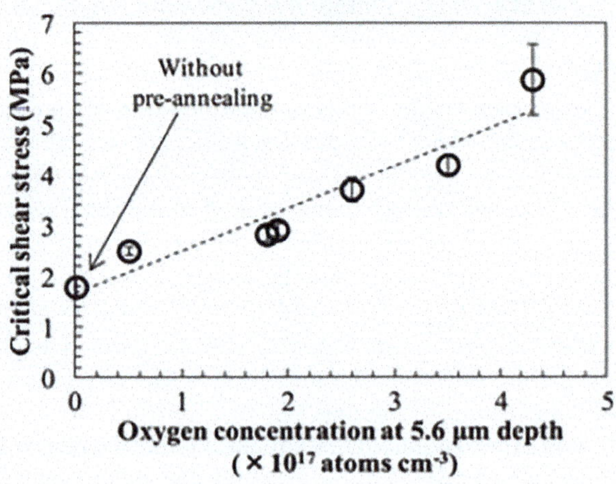

Fig. 1.2 Relationship between critical shear stress and oxygen concentration at a depth of 5.6 μm from the specimen surface. K. J. Fujise, B. Ko, T. Ono and M. Tanaka, ECS Journal of Solid-State Science and Technology, 9, 104,002 (2020). Open access

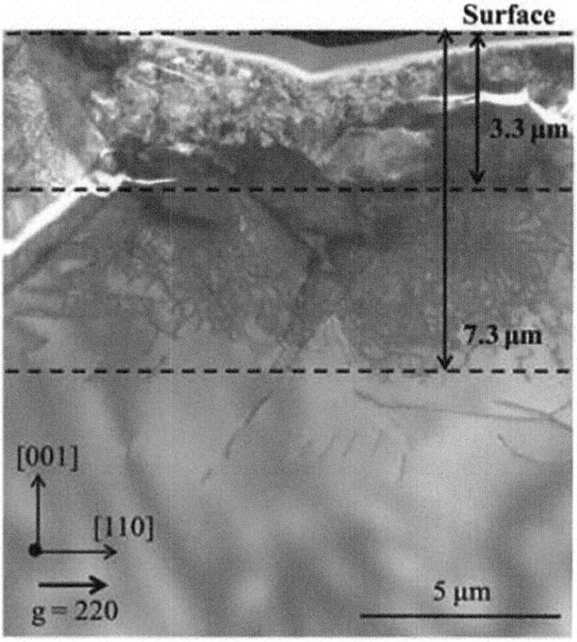

Fig. 1.3 Dark field transmission electron microscopy image displaying the crosssection just beneath the indents (0.98 N) under a two-beam condition. The diffraction vector was set to be g (hkl) = 220. The incident beam direction is [110]. In the image, the range of slip generation depth is 3.3 – 7.3 μm, and the dislocation density decreases as the depth increases. K. J. Fujise, B. Ko, T. Ono and M. Tanaka, ECS Journal of Solid-State Science and Technology, 9, 104,002 (2020). Open access

below the surface by pre-annealing, could depress dislocation motion and thus reducing slip by increasing low locking stress. Wafers with low oxygen concentrations have low locking stress, so slip dislocation can easily be extended. Therefore, on the basis of using these results, it is possible to reduce the risk of wafer bending and leakage current due to slip during manufacturing.

1.1.2 In SiC

Fully dense silicon carbide ceramics were produced by spark plasma sintering (SPS) technique at 1900°C under an applied pressure of 50MPa for 15 min holding time with boron carbide and carbon additions (sintering additives). Oxygen affects all the mechanical properties of the SiC. Microstructure, hardness, and fracture behavior of SPS SiC is presented below as affected by the oxygen content. High oxygen content is detrimental to the final density and hardness. Recall that SiC is an extensively used structural ceramics because its exceptional thermal and mechanical properties, among them high hardness and elastic modulus, oxidation, and wear resistance low of thermal expansion while the density is low.

SG-SiC microstructural images can be seen in Fig. 1.4 made of powders corresponding to 0.22 and 0.38 wt. %. In specimens with 0.38 wt. % oxygen, exaggerated grain growth was observed, while specimens containing 0.22 wt. % oxygen remain with equiaxed and smaller grains and also specimens containing 0.98 wt. % have similar microstructure. By increasing the oxygen content in the powder, the grain size tends to get smaller. The morphology of fracture surface is seen in the microstructures of Fig. 1.5. The fracture modes of all samples are a combination of intergranular and transgranular fracture. Smaller grains appear to show more intergranular fracture while the transgranular mode can be seen through the larger grains. The mixed powder of SiC by ball milling in ethanol and the added C content is seen in Table 1.2, while the densities and elastic modulus (E) of the SG-AW-SiC samples are shown in Table 1.3. It can be clearly seen that the oxygen content affects these properties. Thus, 0.22 wt.% and 0.38 wt.% oxygen content samples densities are very close to theoretical density, however by increasing the oxygen content, the densities, and elastic modulus decrease. The Knoop hardness values are shown in Table 1.4. As seen in the table, by increasing the load the hardness values decrease. In all samples, hardness values follow the same trend.

In summary, dense SiC was produced with addition of B4C and C sintering agents by SPS at 1900°C for 15 min under argon atmosphere. Oxygen content affects all the measured mechanical properties and the microstructure. The hardness results show that high oxygen content is disadvantageous to Knoop hardness values. In the highest oxygen content samples, porosity and softer oxide secondary phases decrease hardness.

1.2 Effect of Moisture

1.2.1 In Si

As an example, on the effect of humidity on the degradation of single crystal silicon microbeam the fatigue strength can be indicated. It was found that the fatigue strength was markedly decreased under a highly humid condition. Scanning electron microscopy (SEM) observation of the fracture surface of a specimen revealed that the morphologies of the fracture surface depended on the applied stress and environmental conditions such as temperature, humidity etc. Understanding the mechanical degradation of a structural material in a MEMS device is crucial for reliable use and commercial success. As long as silicon is protected from the surrounding environment by sealing in a high-vacuum region, silicon shows long-term stability under resonant condition. Fatigue fracture under cyclic loading is accelerated by the existence of moisture for both single-crystal and polycrystal silicon. The test consisted of four series at two different temperatures and two different humidity conditions. Test conditions are shown in Table 1.5. Fatigue properties under different environmental conditions are shown in Fig. 1.6, comparing the effect of humidity. At a high cycle range over 10^8 cycles, the fatigue strength increases with decreasing humidity (See Fig. 6b).

In summary, fatigue tests of single-crystal silicon using a MEMS resonator were performed at different temperatures and humidity conditions. The measurement

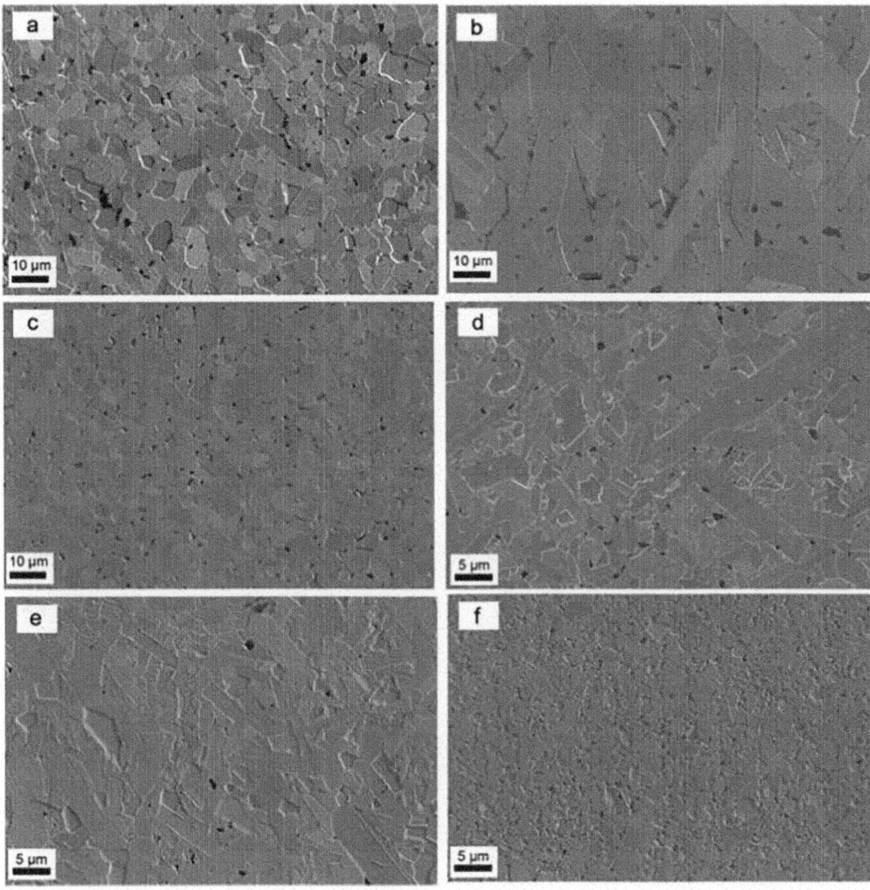

Fig. 1.4 Microstructural images of sample made with SG-SiC **a** 0.22 wt.% oxygen, **b** 0.38wt.% oxygen, **c** 0.98 wt.% oxygen, HC-SiC **d** 0.66 wt.% oxygen, **e** 1.69 wt.% oxygen, and(f) 3.36 wt.% oxygen content powders. Z. A. Yaşar, V. A. DeLucca, and R. A. Haber, Ceramics International, 44, 23,248 (2018). With kind permission of Elsevier

results demonstrate that humidity has a large effect on fatigue strength and fatigue life, particularly at low temperature. The morphology of the fracture surface was only related to the applied maximum stress and of no relation to environmental conditions.

1.2.2 In SiC

SiC exhibits high hardness and elastic modulus, low thermal expansion, and good thermal stability. At ambient temperature, SiC ceramics are exceptionally stable and do not react with surrounding atmosphere. However, when two ceramics slide against

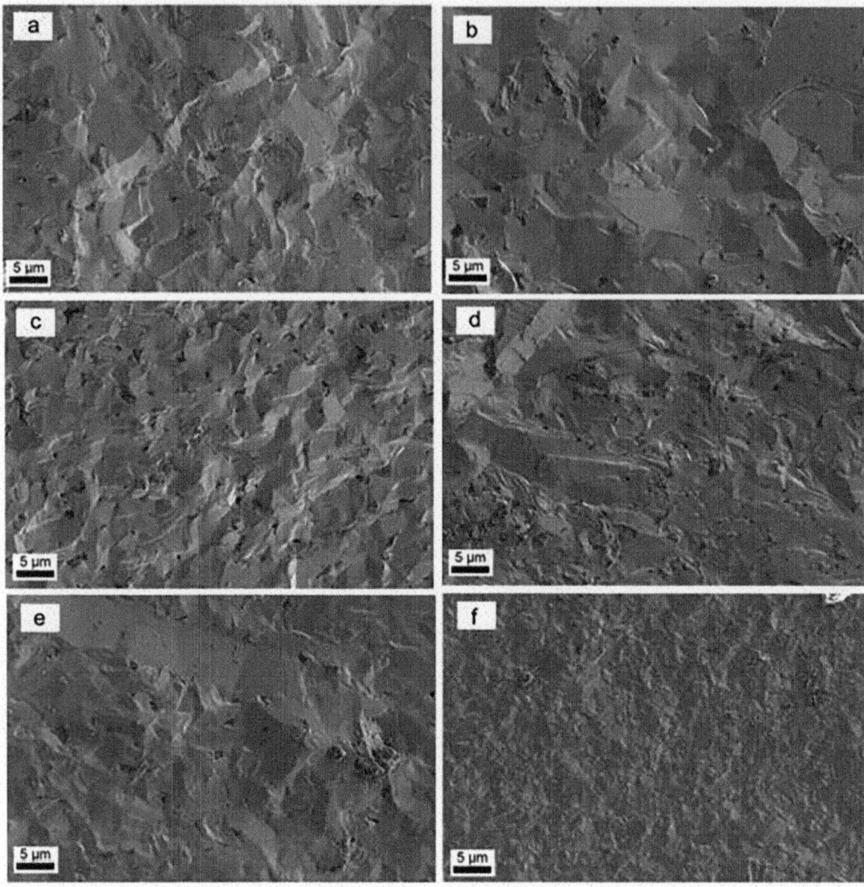

Fig. 1.5 Fracture surface images of SG-AW series with series **a** 0.22 wt.% oxygen, **b** 0.38 wt.% oxygen, **c** 0.98 wt.% oxygen, HC-SiC **d** 0.66 wt.% oxygen, **e** 1.69 wt.% oxygen, and **f** 3.36 wt.% oxygen content. Z. A. Yaşar, V. A. DeLucca, and R. A. Haber, Ceramics International, 44, 23,248 (2018). With kind permission of Elsevier

Table 1.2 Silicon carbide samples matrix. Z. A. Yaşar, V. A. DeLucca, and R. A. Haber, Ceramics International, 44, 23,248 (2018). With kind permission of Elsevier

Sample series	Sample	Oxygen content wt.%	Carbon added wt.%
SG- AW-series	Saint Gobain	**0.22**	1.0
SG- AW-series	Saint Gobain	0.38	1.0
SG- AW-series	Saint Gobain	0.98	1.0
HC-AW-series	HC Starck	**0.66**	1.5
HC-AW-series	HC Starck	1.69	1.5
HC-AW-series	HC Starck	3.36	1.5

Table 1.3 SG-AW and HC-AW series mechanical and microstructural properties. Z. A. Yaşar, V. A. DeLucca, and R. A. Haber, Ceramics International, 44, 23,248 (2018). With kind permission of Elsevier

Sample (%)	Density(g/cni3)	Average grain size (std.dev.)	Phase fraction 4H (%)	Phase fraction 6H (%)	4II/6II ratio	E(GPa)
SG-AW-0.22	3.19	5.43 (0.89)	17	71	0.24	440
SG-AW-0.38	3.18	13.8 (4.19)	61	31	1.97	437
SG-AW-0.98	3.12	4.18 (0.88)	16	63	0.25	434
HC-AW-0.6	3.21	2.89 (0.99)	30	53	0.57	435
HC-AW-1.69	3.20	2.28 (1.91)	23	53	0.43	453
HC-AW-3.36	3.15	1.21 (0.24)	12	24	0.46	405

Table 1.4 SG-AW series and HC-AW series Knoop hardness results. Z. A. Yaşar, V. A. DeLucca, and R. A. Haber, Ceramics International, 44, 23,248 (2018). With kind permission of Elsevier

Knoop hardness (GPa)					
Sample (%)	100 (g)	300 (g)	500 (g)	1000 (g)	2000(g)
SG-AW-0.22	30.11 ± 1.25	24.47 ± 0.66	22.41 ± 0.64	21.98 ± 0.88	21.38 ± 0.19
SG-AW-0.38	27.01 ± 1.58	23.20 ± 0.62	22.20 ± 0.61	21.53 ± 0.94	21.10 ± 0.99
SG-AW-O.98	28.66 ± 0.79	23.90 ± 0.63	22.27 ± 0.62	21.11 ± 0.38	20.66 ± 0.41
HC-AW-0.60	28.38 ± 1.33	24.26 ± 1.16	22.75 ± 0.43	21.63 ± 0.41	21.09 ± 0.34
HC-AW-1.69	28.59 + 1.99	23.98 + 0.56	22.96 ± 0.58	21.21 ± 0.54	21.04 ± 0.93
HC-AW-3J6	26.93 ± 1.50	22.14 ± 0.62	20.96 ± 0.43	19.90 ± 0.23	19.17 ± 0.37

Table 1.5 Environmental conditions in fatigue test. T. Ando, M. Shikida and K. Sato, Sensors, and Materials, 28, 113 (2016). Open access

Temperature (°C)	25		70	
Relative humidity (RH) (%)	30	85	0.5	15
Absolute humidity (g/m^3)	6.198	19.60	0.9861	29.58

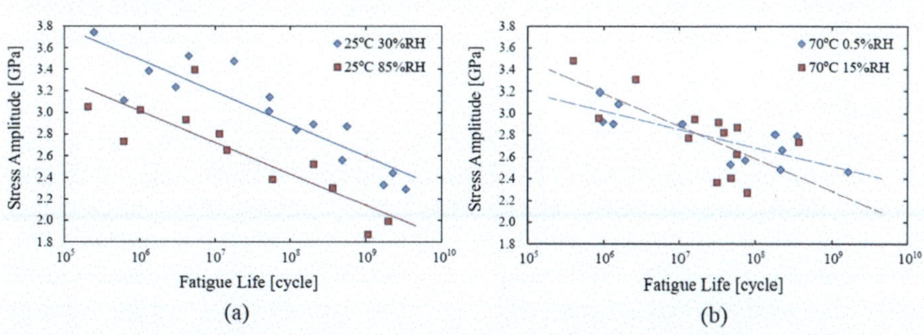

Fig. 1.6 S–N plots of fatigue testing results in various environments at **a** 25 and **b** 70°C. T. Ando, M. Shikida and K. Sato, Sensors, and Materials, 28, 113 (2016). Open access

each other, mechanical processes such as indentation, abrasion and attrition produce a large number of tiny debris and the surface area of a material increases by several orders of magnitude. Besides processes such as triboemission, generation of lattice defects, intense heat and phase transformation aid these reaction(s) and accelerate the overall kinetics. These stimulated reactions are widely referred to as 'tribochemical'. SiC is an important engineering structural material and is widely used in applications like bearings, mechanical seals, and cylinder liners. Therefore, friction and wear properties of SiC are important to investigate. During experimentation, various parameters were monitored with variation of humidity (Table 1.7).

A pin-on-disc tester was used to study the dry sliding wear behavior of different SiC tribo-pairs. All tests were conducted in air (relative humidity levels varying from 30 to 60%) for a sliding distance of 130 m maintaining a sliding speed of 36 mm/s (100 rpm) with normal load of 1N. The ball (8 mm diameter) material was a commercially available α-SiC (mainly 6H) with small amounts of B (3700 ppm) and C (not known), which were added to densify the material to 99.3%. Note, metal oxides, Al-B-C, AlN–metal oxides as sintering additives during liquid-phase sintering tailor the microstructure and improve fracture toughness of SiC ceramics. The average grain size of the ball SiC was ~ 7 μm. The mechanical properties of the ball material are given in Table 1.6, while in Fig. 1.7 the mechanical properties of β-SiC disc materials with different dopants is presented. The average grain size of disc samples is ~ 1 μm. Typical microstructures of SiC materials that were used for balls as well as discs are shown in Fig. 1.7. Both microstructures consisted of carbon (as graphite) in different forms. In the ball material, the graphite is present as small, isolated particles either at the triple junctions or at the grain boundaries. In the SiC disc materials, graphite exists as long needles having an average thickness < 1.0 μm and the length varies from 5 to 35 μm as illustrated in Fig. 7b. The variation of the coefficient of friction (COF) as a function of humidity and doping elements is shown in Fig. 1.9. At high relative humidity (RH: 60%), the variation in COF values is minimal. The largest variation in the friction was at the lower humidity but depends on the dopant. Specific wear rates of ball and disc in undoped and doped samples with variation of humidity is shown in Table 1.8 and in Fig. 1.8. The differences in wear rates are reflected in the microstructural features (Murthy et al. 2004). In SiC, both COF and wear rate decrease with increasing humidity as seen in Table 1.8 and Figs.

8 and 9.

In summary, the effects of relative humidity and doping elements on the friction and wear properties of SiC in unlubricated condition was presented. At higher humidity, the friction coefficient values are low and independent of doping elements. The specific wear rate is lower at the higher humidity (60% RH) as compared to the lower humidity (i.e., at 30% RH) values. As the relative humidity increases from 30 to 60%, the wear mechanism gradually changes from mechanical to tribochemical. Detail of the experimental and the results are extensively described and discussed in Murthy et al. (Murthy et al. 2004). It might be noted though that the tribo-layer consisting of carbon and hydrated silica progressively smears with relative humidity and becomes instrumental in controlling the friction and wear properties.

Table 1.6 Mechanical properties of α-SiC ball material with 3700 ppm of B. V. S. R. Murthy, H. Kobayashi, S. Tsurekawa, N. Tamari, T. Watanabe and K. Kato, Tribology International, 37, 353 (2004). With kind permission of Elsevier

Property	
Density (% theoretical)	99.3
Hardness (GPa)	23
Fracture toughness (MPa Vm)	2.2
Elastic modulus (GPa)	410
Flexural strength (MPa)	550

Table 1.7 Mechanical properties of β-SiC disc materials with different dopants. V. S. R. Murthy, H. Kobayashi, S. Tsurekawa, N. Tamari, T. Watanabe and K. Kato, Tribology International, 37, 353 (2004). With kind permission of Elsevier

Material	Dopant (ppm)	Density (% of theoretical)	Average grain size (pm)	Average hardness (GPa)	Average fracture toughness (MPa Vm)	Average flexural strength[a] (MPa)
SiC	–	98.9	1.15	28	3.3	1050
SiC	Al: 2100	99.1	1.10	24	3.5	760
SiC	Mg: 3100	99.0	2.50	24	3.8	815
SiC	P: 440	99.4	1.12	28	3.6	820

[a]Four-point bend test

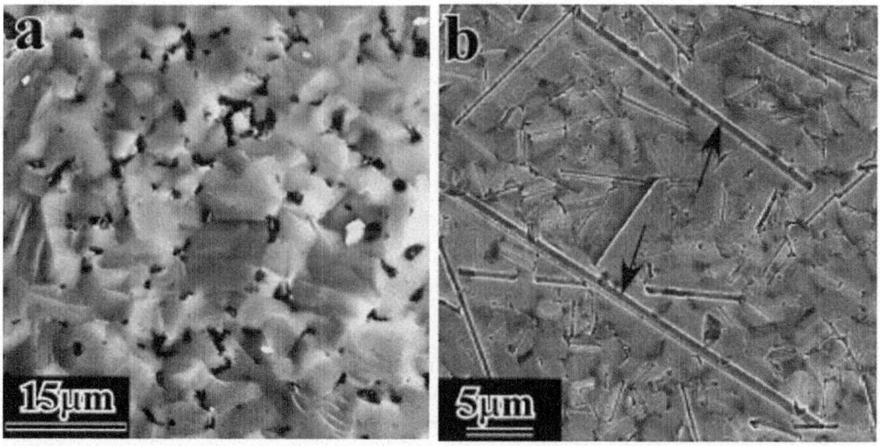

Fig. 1.7 **a** Fracture surface of a α-SiC ball showing the grain size and distribution of graphite as isolated black particles which are smaller than about 1 μm. **b** Etched microstructure of Mg-doped β-SiC disc showing the locations where long needles of graphite (thinner than about 1 μm) appeared. The locations are highlighted by arrows. V. S. R. Murthy, H. Kobayashi, S. Tsurekawa, N. Tamari, T. Watanabe and K. Kato, Tribology International, 37, 353 (2004). With kind permission of Elsevier

Table 1.8 Specific wear rates of ball and disc in undoped and doped samples with variation of humidity. V. S. R. Murthy, H. Kobayashi, S. Tsurekawa, N. Tamari, T. Watanabe and K. Kato, Tribology International, 37, 353 (2004). With kind permission of Elsevier

Doping elements				
Relative humidity (%)	Al	Mg	P	No dopant
Specific wear rate of ball $W_{sb} \times 10^{-6}$ mm^3/Nm				
30	3.99	0.31	0.34	2.11
45	0.45	0.66	0.18	0.16
60	0.11	0.02	0.06	0.07
Specific wear rate of ball $W_{sd} \times 10^{-6}$ mm^3/Nm				
30	1.05	0.37	0.40	0.67
45	0.28	0.74	0.29	0.26
60	0.31	0.34	0.27	0.18

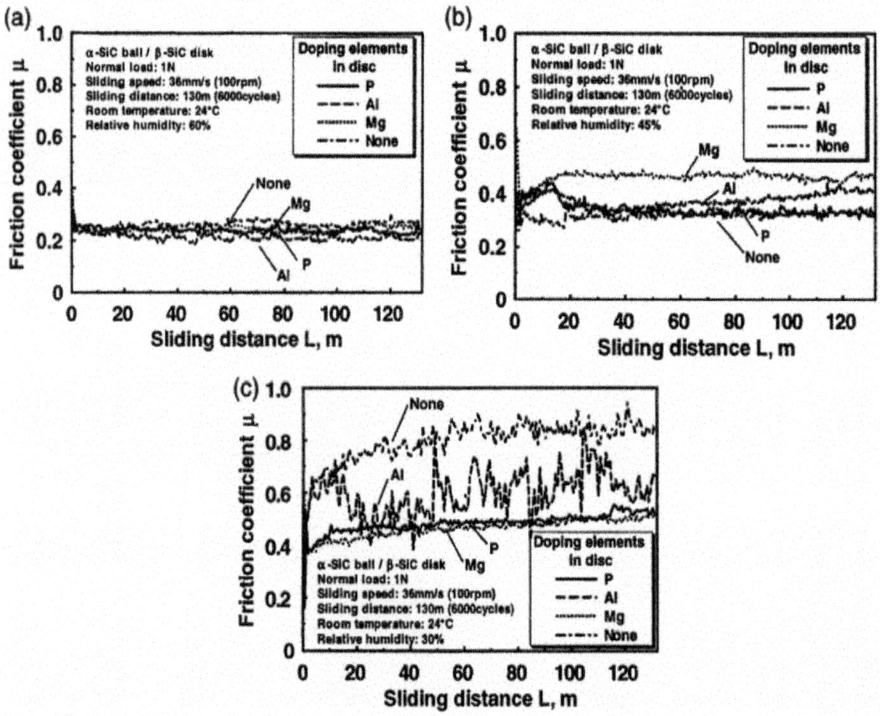

Fig. 1.8 The combined effect of relative humidity and doping elements on the coefficient of friction **a** at 60% RH, **b** 45% RH and **c** 30% RH. V. S. R. Murthy, H. Kobayashi, S. Tsurekawa, N. Tamari, T. Watanabe and K. Kato, Tribology International, 37, 353 (2004). With kind permission of Elsevier

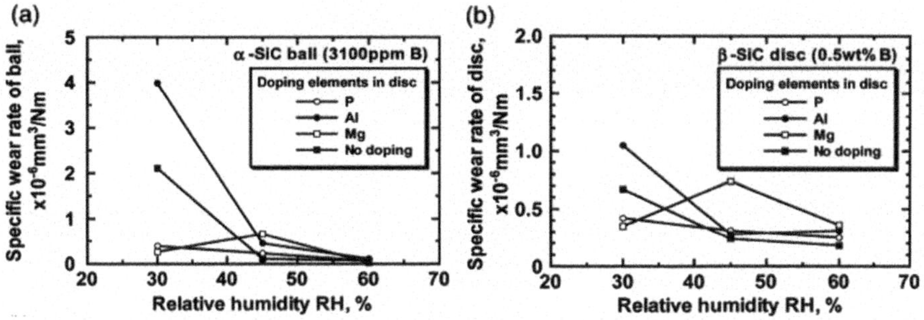

Fig. 1.9 **a** Variation of specific wear rate of α-SiC when slid against different β-SiC disc samples. **b** Specific wear rate of different β-SiC discs (doped as well as undoped) under similar sliding conditions. The wear rate of ball is higher than the disc samples. Note the reduction in the specific wear rate with relative humidity content. V. S. R. Murthy, H. Kobayashi, S. Tsurekawa, N. Tamari, T. Watanabe and K. Kato, Tribology International, 37, 353 (2004). With kind permission of Elsevier

1.2.3 In GaAs

Micromechanical testing techniques to measure the mechanical properties of material at the microscopic scale has become a primary interest for a wide range of applications, such as thin film technology, microelectromechanical systems (MEMS) and composite materials. The mechanical properties of micro- and nanoscale components can differ significantly from bulk material, as they are affected by factors like fabrication process, crystal size, dimensional constraints, and surface characteristics.

All single-crystal GaAs microtensile specimens tested by this method showed brittle failure and except one specimen (which was discarded) failed within the gauge section. Stress and strain data for tensile experiments are illustrated in 1.10 (f). The microtensile samples exhibited ultimate failure stress of 1.86 ± 0.17 GPa and ultimate strains of 3.0 ± 0.4%. The elastic modulus in the [001] crystal orientation, obtained from the unloading is displayed in Fig. 10c. The measured yield stress was 2.60 ±

0.14 GPa, and the yield strain was 3.5 ± 0.3%. The apparent elastic modulus measured on the [001] orientation was 86.5 ± 3.1 GPa. After yielding, all micropillars exhibited a consistent and reproducible plastic regime at 2.42 ± 0.11 GPa, with no apparent hardening. Micropillars showed a clear stress drop of a few hundred MPa before reaching the constant stress regime. STEM imaging performed on a micropillar compressed to 5.1% strain, seen in Fig. 11b, showed two main contrast bands and the presence of multiple dislocations. The first contrast band was oriented perpendicularly to the loading direction, while the second was oblique at an angle of 36.3° with respect to the pillar [001] axis and had a thickness of approximately 80 nm.

A surface step was observed in the amorphous layer of the pillar at the end of the second contrast band. TKD analysis, observed in Fig. 11a established that these bands correspond to crystal twinning in two different {111} planes. Assuming a perfect

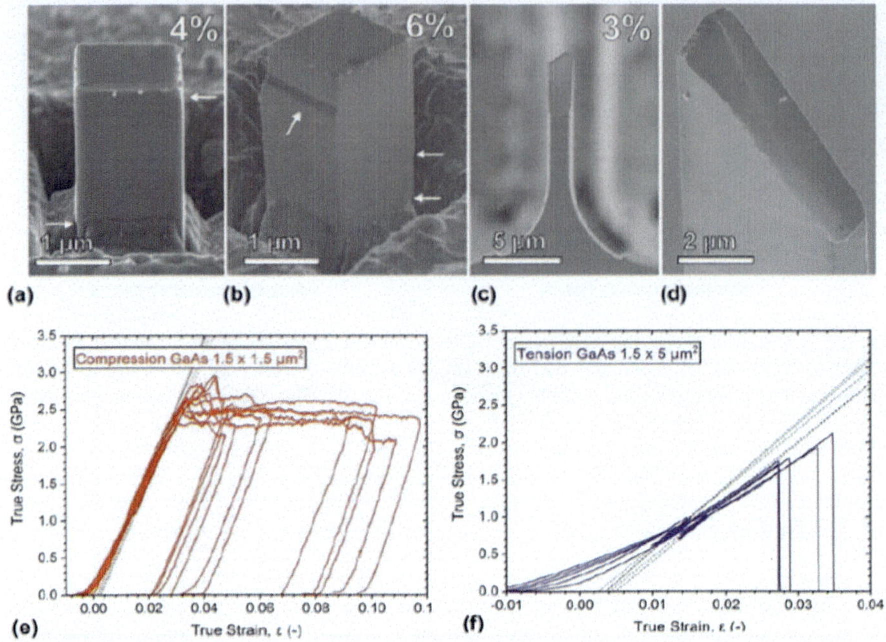

Fig. 1.10 **a, b** SEM images of two GaAs micropillars tested in compression and **c, d** the fracture surface of a microtensile specimen after brittle failure. White arrows indicate visible surface steps. **e, f** Stress–strain data for micropillar compression **e** and microtensile tests **f** on GaAs with the [100] direction of the crystal aligned with the loading axis. Dotted lines denote the extension of the unloading slope, where elastic modulus was measured. D. Casari, L. Pethö, P. Schürch, X. Maeder, L. Philippe, J. Michler, P. Zysset and J. Schwiedrzik, Journal of Materials Research, 34, 2517(2019). Open access

alignment between the [001] axis and the loading axis, the resolved shear stress on the slope at around 1.5% strain, was 81.4 ± 4.6 GPa. SEM images of GaAs micropillars tested are shown in Fig. 1.10. Uniaxial compression tests on taper-free single-crystal GaAs micropillars revealed close to perfect plasticity. The experimental data are {111} planes were found to be 1.0 ± 0.1 GPa. TKD stands for Transmission Kikuchi Diffraction.

Single-crystal GaAs has been mechanically tested along the [001] crystallographic in GaAs needs to occur at sizes between 1.7 and 2.3 μm. In contrast, brittle-to-ductile transition in tension was not observed, but it is expected at lower length scales. Plasticity in compression was attributed to twinning formation along multiple {111} planes.

Single-crystal GaAs has been mechanically tested along the [001] crystallographic orientation in both compression and tension on samples sharing a common thinnest dimension of 1.70 ± 0.19 μm. The elastic modulus in both compression and tension was measured and was in good agreements with the value found in literature of 85.5 GPa. While the elastic response was the same in tension and in compression, all tensile

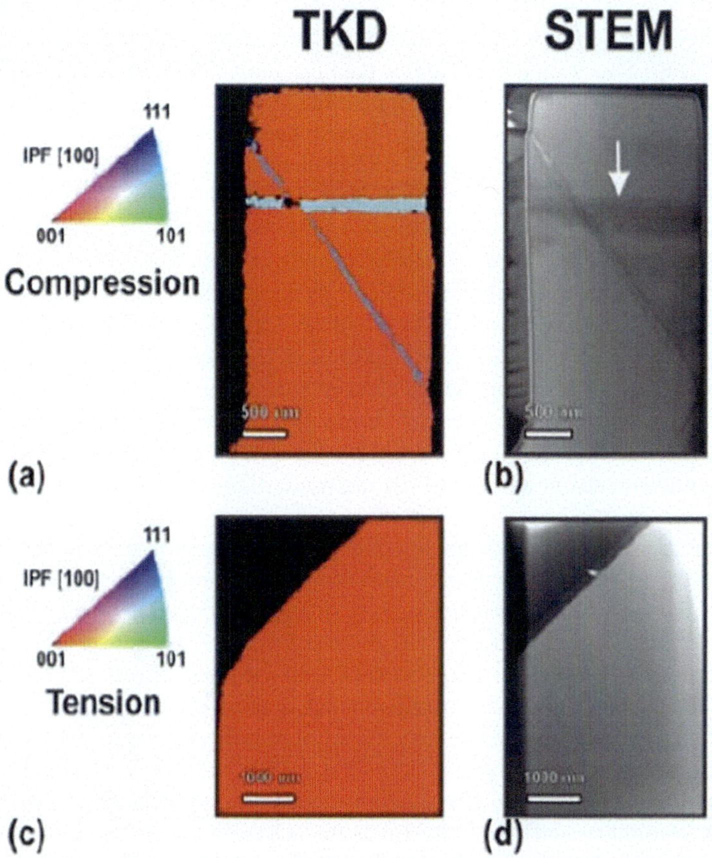

Fig. 1.11 **a** TKD and **b** STEM images for a compressed micropillar. Twinning in the {111} system was identified as the major deformation mechanism by TKD. High dislocation density may be observed near twin boundaries based on bright field STEM (white arrow) **c** TKD and **d** STEM of the lower portion of a failed microtensile sample. No measurable dislocation activity was observed by either TKD or STEM in the case of tensile loading. D. Casari, L. Pethö, P. Schürch, X. Maeder, L. Philippe, J. Michler, P. Zysset and J. Schwiedrzik, Journal of Materials Research, 34, 2517(2019). Open access

samples exhibited brittle failure, whereas all micropillars deformed plastically. At the macroscale, GaAs is typically known to have very little room temperature ductility. Brittle-to-ductile transition under uniaxial compression at room temperature has been reported for micropillars of diameters in the range of 1 μm. By combining the results of this study with previous observations it can be concluded that for compression, the brittle-to-ductile transition in tension was not observed, but it is expected at lower length scales. Plasticity in compression was attributed to twinning formation along multiple {111} planes.

GaAs, compression along the [001] axis generates a total of 8 slip systems {111} < 110 > having a Schmidt factor of 0.408. A dislocation of the indicated systems can dissociate into Shockley partials and a stacking fault (SF) in the {111} glide plane. The dissociation distance at room temperature in equilibrium is about 613 nm and can be influenced by external stresses. The SF can act as a preferred site for another dislocation nucleation on an adjacent sliding plane and can lead to the formation of a lateral growth twin. In taper-free micropillars a constant plastic flow (for strain up to 12% at 2.42.GPA) was observed without hardening, which suggests that nucleated dislocations escape the pillar. In an almost pure crystal, plasticity is initiated only when a certain number of mobile dislocations are available.

Tensile tests showed brittle failure and no sign of dislocation nucleation was observed (Fig. 1.11). It was noticed that most of the edges of the fractured surfaces coincided with a {110} plane, which correspond to the cleavage plane for materials with a zinc blende structure, such as GaAs. The complicated shape of the fracture surface observed in Fig. 10d is caused by the competition of the {100} plane of maximum tensile stress and the {110} cleavage plane, on which cracks can easily propagate in GaAs.

In summary, a methos allowing uniaxial tensile testing is described in this section. The micro tensile method was validated on (100) Si and by experimentally evaluating the tension–compression asymmetry of strength in (100) single-crystal GaAs. Clear loading mode dependence was observed when tensile tests were compared with micropillar compression experiments. GaAs showed plasticity via twinning in compression, while failing in a brittle manner in tension. A size effect is observed. GaAs is stronger in compression, reaching a plastic flow stress of 2.42 ± 0.11 GPa, while it fails in tension at 1.86 ± 0.17 GPa. STEM and TKD were used to assess deformation mechanisms in both loading modes. The experimental results showed high reproducibility and reliability for both measured elastic modulus and tensile strength, showing the potential of the novel microtensile testing methodology.

References

T. Ando, M. Shikida, K. Sato, Sensors and Materials **28**, 113 (2016)

D. Casari, L. Pethö, P. Schürch, X. Maeder, L. Philippe, J. Michler, P. Zysset, J. Schwiedrzik, J. Mater. Res. **34**, 2517 (2019)

J. Fujise, B. Ko, T. Ono, M. Tanaka, ECS J. Solid State Sci. Technol. **9**, 104002 (2020)

V.S.R. Murthy, H. Kobayashi, S. Tsurekawa, N. Tamari, T. Watanabe and K. Kato. Tribol. Int. **37**, 353 (2004)

Z.A. Yaşar, V.A. DeLucca, R.A. Haber, Ceram. Int. **44**, 23248 (2018)

Chapter 2
Elastic Properties

2.1 The Effect of Oxygen

2.1.1 In SiC

Chapter 2 considers elastic properties with yielding as the demarcation point between elasticity and plasticity. Ultrasound analysis results showed that the high oxygen content affected the elastic properties. SiC samples with the high oxygen content had lower elastic moduli, shear moduli and bulk moduli. It is clear that increasing the oxygen content decreases the elastic properties.

Two types of SiC powders were treated by HF etching, aging and heat treating tp produce various oxygen content in the samples. Spark plasma sintering (SPS) at 1900°C for 15 min holding time was applied at a pressure of 50 MPa to obtain the specimens. Boron carbide and carbon additions were added as sintering agents. Microstructure, density Knoop hardness and fracture behavior of SiC, were examined to understand the effect of oxygen content on the properties of dense silicon carbide. The results show that oxygen content affects all of the measured mechanical and microstructural properties. High oxygen content is shown to be detrimental to the final density and hardness of SPS SiC. As known, SiC is a structural ceramic due to its exceptional thermal and mechanical properties such as high hardness, high elastic modulus, good wear and oxidation resistance, high thermal conductivity, low coefficient of thermal expansion and low density (3.21 g/cm^3). Because its properties, dense SiC is a good candidate for use in high temperature, automotive, corrosion resistant applications, wear components and ceramic armor Yasar et al. (2018).

B and C were used as sintering additives. The Silicon carbide samples are listed in Table 2.1. After sintering, the dense samples were sandblasted to remove the layer of C (graphite) foil from SPS. Surface grinding was necessary to produce a smooth and flat surface for ultrasound analysis which measures the elastic modulus. The linear intercepts method was used to determine average grain size, and at least 100

J. Pelleg, *The Influence of Environment on Mechanical Characteristics of Key Semiconductor Components*, https://doi.org/10.1007/978-3-031-34530-2_2

Table 2.1 Silicon carbide samples matrix. Z. A. Yasar, V. A. DeLucca and R. A. Haber, Ceramics International, 44, 23,248 (2018). With kind permission of Elsevier

Sample series	Sample	Oxygen content wt%	Carbon added wt%
SG- AW-series	Saint Gobain	0.22	1.0
SG- AW-series	Saint Gobain	0.3896	1.0
SG- AW-series	Saint Gobain	0.9896	1.0
HC-AW-series	HC Starck	0.6696	1.5
HC-AW-series	HC Starck	*1.69*	1.5
HC-AW-series	HC Starck	*3.36*	1.5

intercepts were measured on each micrograph. Knoop hardness measurements at five loads were performed by five indents.

using a microhardness tester. The loads were 100 g, 300 g, 500 g, 1000 g, and 2000 g Fig. 1d–f. Microstructural images can be seen in Fig. 1a–c, while performed by five indents using a microhardness tester. By increasing the oxygen content in the powder, the grain size tends to get smaller. The 3.36 wt% oxygen sample shows a lower density than the others, and also displays small and equiaxed grain shape. This specimen is shown by EDS in Fig. 2.2. Further, Electron Backscatter Diffraction (EBSD) maps of both the SG-SiC (a-c) and HC-SiC (d-f) samples. are shown in Fig. 2.3. The densities and elastic modulus (E) of the SG-AWSiC samples are shown in Table 2.2. It can be clearly seen that the oxygen content affects these properties. 0.22 wt% and 0.38 wt% oxygen content samples densities are very close to theoretical density, however by increasing the oxygen content, the densities, and E modulus decrease. SG-AW- series Knoop hardness-load results are shown in Table 2.3. As expected, hardness values decrease with increasing load. All samples hardness values follow the same trend. At lower load (100 g, 300 g) the 0.38 wt% oxygen content sample shows lower hardness than other samples. This might be a grain size effect since this sample has a larger average grain size (13.8 μm). At higher loads (500 g, 1000 g, and 2000 g), the hardness decreases with increasing oxygen content like the density and elastic modulus values. The HC-AW series Knoop hardness results are shown in Table 2.4 for the decreased hardness. The lowest oxygen content samples have a higher hardness than the others. The AW and HC-AW series samples show mixed-mode fracture behavior, with large grains tending to have a transgranular mode while small grains display more intergranular fracture mode (Fig. 2.4).

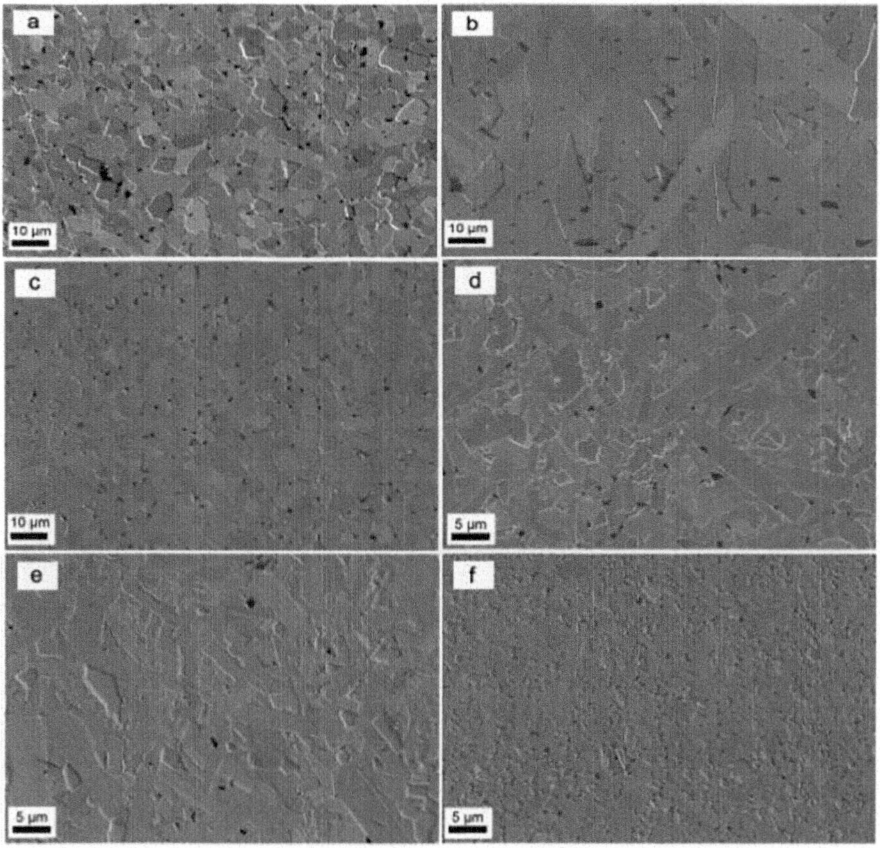

Fig. 2.1 Microstructural images of sample made with SG-SiC **a** 0.22 wt% oxygen, **b** 0.38 wt% oxygen, **c** 0.98 wt% oxygen, HC-SiC **d** 0.66 wt% oxygen, **e** 1.69 wt% oxygen, and **f** 3.36 wt% oxygen content powders. Z. A. Yasar, V. A. DeLucca and R. A. Haber, Ceramics International, 44, 23,248 (2018). With kind permission of Elsevier

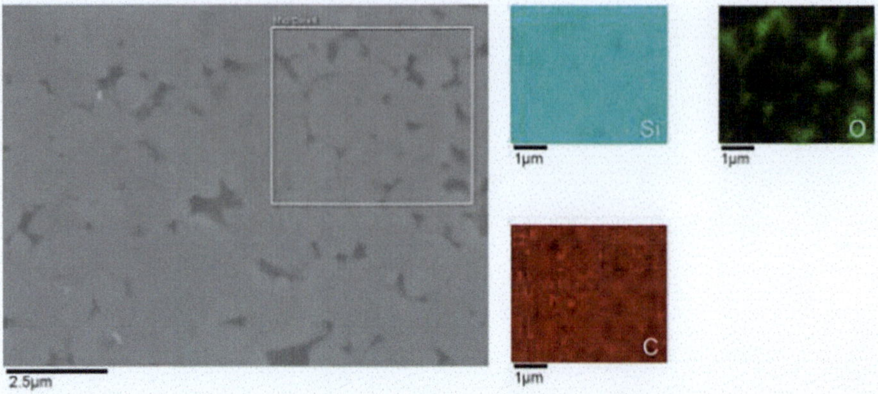

Fig. 2.2 EDS maps of HC-AW-3.36% SiC sample, showing Si in blue, O in green and C in red. Z. A. Yasar, V. A. DeLucca and R. A. Haber, Ceramics International, 44, 23,248 (2018). With kind permission of Elsevier

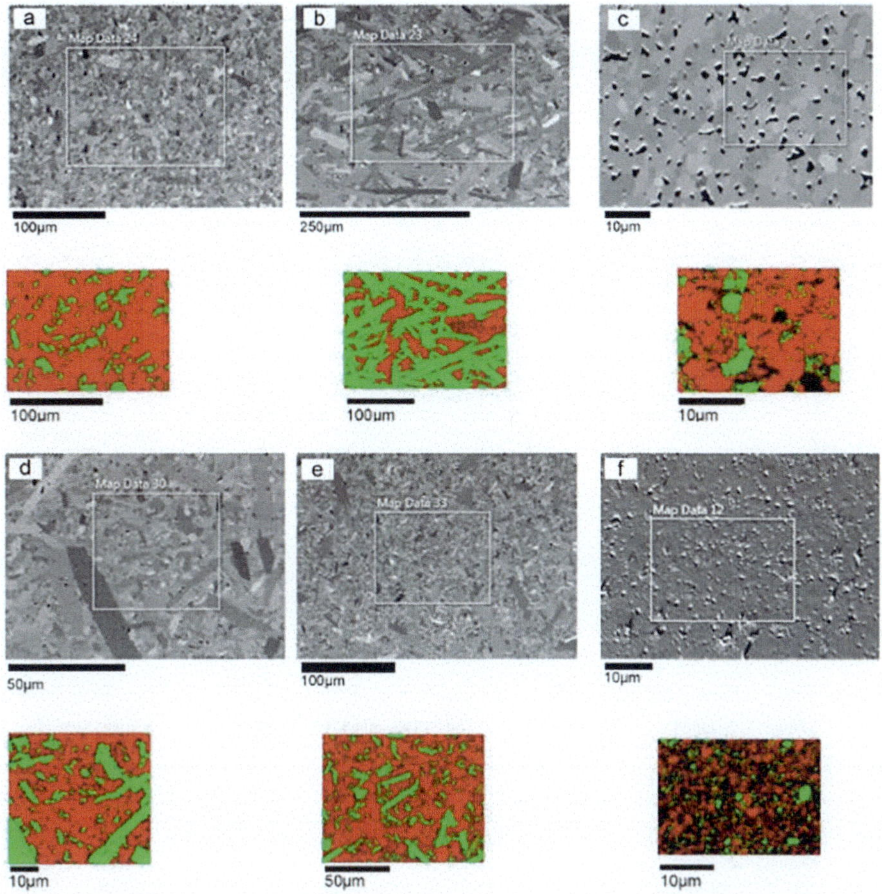

Fig. 2.3 EBSD maps for SG-AW series **a** 0.22 wt% oxygen, **b** 0.38 wt% oxygen, **c** 0.98 wt% oxygen, HC-SiC **d** 0.66 wt% oxygen, **e** 1.69 wt% oxygen, and **f** 3.36 wt% oxygen content, showing 6H-polytype in red, 4H-polytype in green and zero solution in black color. Z. A. Yasar, V. A. DeLucca and R. A. Haber, Ceramics International, 44, 23,248 (2018). With kind permission of Elsevier

Table 2.2 SG-AW and HC-AW series mechanical and microstructural properties. Z. A. Yasar, V. A. DeLucca and R. A. Haber, Ceramics International, 44, 23,248 (2018). With kind permission of Elsevier

Sample	Density (g/ cm^3)	Average grain size (std.dev.)	Phase fraction 4 H (%)	Phase fraction 6 H (%)	4 H/6 H ratio	E (Gpa)
SG-AW-0.22%	3.19	5.43 (0.89)	17	71	0.24	440
SG-AW-0.38%	3.18	13.8 (4.19)	61	31	1.97	437
SG-AW-0.98%	3.12	4.18 (0.88)	16	63	0.25	434
HC- AW-0.6%	3.21	2.89 (0.99)	30	53	0.57	435
HC- AW-1.69%	3.20	2.28 (1.91)	23	53	0.43	453
HC- AW-3.36%	3.15	1.21 (0.24)	12	24	0.46	405

Table 2.3 SG-AW series Knoop hardness results. Z. A. Yasar, V. A. DeLucca and R. A. Haber, Ceramics International, 44, 23,248 (2018). With kind permission of

Sample (%)	Knoop hardness (GPa)				
	100 (g)	300 (g)	500 (g)	1000 (g)	2000 (g)
SG-AW-0.22	30.11 ± 1.25	24.47 ± 0.66	22.41 ± 0.64	21.98 ± 0.88	**21.38± 0.19**
SG-AW-0.38	27.OI ± I.58	23.20 ± 0.62	22.20 ± 0.61	21.53 ± 0.94	21.10 ± 0.99
SG-AW-0.98	28.66 ± **0.79**	23.90 ± **0.63**	22.27 ± **0.62**	21.11 ± **0.38**	20.66 ± **0.41**

Table 2.4 HC-AW series Knoop hardness results. Z. A. Yasar, V. A. DeLucca and R. A. Haber, Ceramics International, 44, 23,248 (2018). With kind permission of

Sample	Knoop hardness (GPa)				
	100 (g)	300 (g)	500 (g)	1000 (g)	2000 (g)
HC-AW-0.60%	28.38 ± 1.33	24.26 ± 1.16	22.75 ± 0–43	21.63 ± 0.41	21.09 ± o.34
HC-AW-1.69%	28.59 ± 1.99	23.98 ± 0.56	22.96 ± 0.58	21.21 ± 0.54	21.04 ± o.93
HC-AW-3.36%	26.93 ± 1.50	22.14 ± 0.62	20.96 ± 0.43	19.90 ± 0.23	19.17 ± 0.37

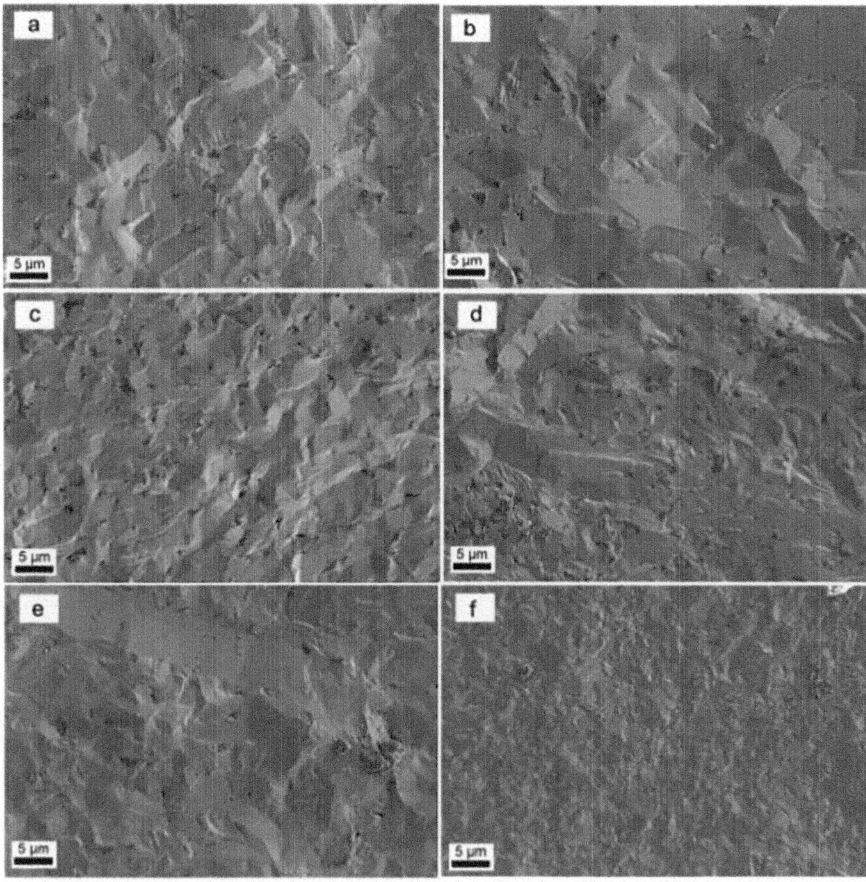

Fig. 2.4 Fracture surface images of SG-AW series with series **a** 0.22 wt% oxygen, **b** 0.38 wt% oxygen, **c** 0.98 wt% oxygen, HC-SiC **d** 0.66 wt% oxygen, **e** 1.69 wt% oxygen, and **f** 3.36 wt% oxygen content. Z. A. Yasar, V. A. DeLucca and R. A. Haber, Ceramics International, 44, 23,248 (2018). With kind permission of Elsevier

Reference

Z.A. Yaşar, V.A. DeLucca, R.A. Haber, Ceramics International **44**, 23248 (2018)

Chapter 3
Plastic Deformation

3.1 The Effect of Oxygen

3.1.1 In Si

The effect of oxygen on plastic deformation in single crystal silicon is presented in this section. The mechanism of plasticity in silicon is complex and depends on many factors among them on loading conditions, dislocations, phase transformation and chemical reactions. In general plastic deformation in silicon is the coupled result controlled by the stress field applied, chemical interaction governed by both the loading type and environment. Thus, increase in temperature accelerates penetration of oxygen into silicon and reduces the critical stress of yielding. Plasticity of silicon in the form of phase transformation (from amorphous to bcc structure) is determined by loading history.

Components in micro-electro-mechanical systems require high surface finish and damage-free subsurface, therefore deformation mechanisms in the material during surface processing must be fully understood. At ambient temperature single crystal silicon is brittle, however, its amorphous modification exhibits a plastic flow under low loads so that possible phase transformation events in machining operation are of great significance in producing damage-free components by machining (Zhang and Zarudi 2001).

The subsurface structure after loading was investigated by TEM and scanning transmission electron microscope (STEM) by different techniques (Zhang and Zarudi 2001). The subsurface structures of silicon after indentation, scratching, grinding, and polishing are explored through Figs. 3.1, 3.2, 3.3 and 3.4, respectively. A common feature of the structures is the phase transformations in all the specimens subjected to different loading conditions.

The shape of the indentation induced transformation zone depends on the magnitude of the load. At a low load of 390–40 mN almost all the area is amorphous as seen in Fig. 3.1a. Beyond 50 mN particles of BCC silicon appear in the middle of

J. Pelleg, *The Influence of Environment on Mechanical Characteristics of Key Semiconductor Components*, https://doi.org/10.1007/978-3-031-34530-2_3

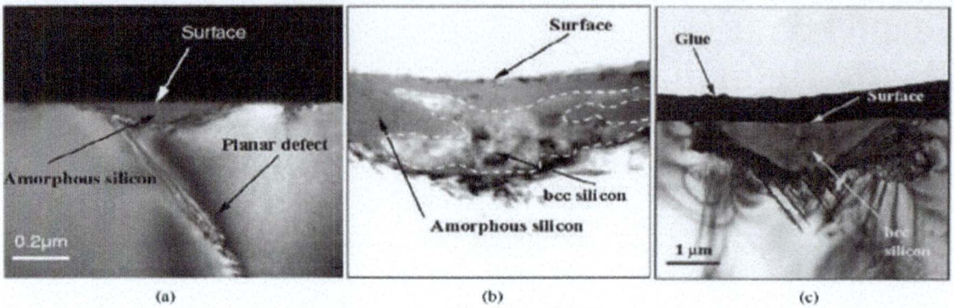

Fig. 3.1 The cross-sectional view of indentation-induced deformation in the subsurface of silicon specimens after complete unloading. Note the bcc-silicon particles with lighter colors, highlighted by the dotted white curves in (b). **a** 30 mN, **b** 70 mN, and **c** 90 mN. L. Zhang and I. Zarudi, International Journal of Mechanical Sciences 43, 1985 (2001). With kind permission of Elsevier

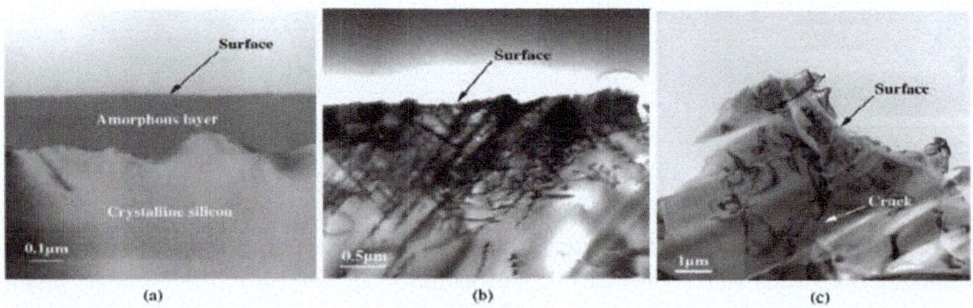

Fig. 3.2 A cross-sectional view of the scratching-induced deformation in the subsurface of the silicon specimens under various loads. Note the coexistence of dislocations and planar defects in (b). In both the cases shown in (b) and (c), amorphous layers on the scratched surfaces dislodged due to micro-cracking and can no longer be viewed on the photos: **a** 4.4 N, **b** 7.8 N, and **c** 11.2 N. L. Zhang and I. Zarudi, International Journal of Mechanical Sciences 43, 1985 (2001). With kind permission of Elsevier

the zone as seen in Fig. 3.1b which growth quickly with increased load. It occupies almost all the transformation zone at 90 mN as illustrated in Fig. 3.1c. During the deformation planar defects appear, but no dislocations. Micro-cracking is observed at a load of 100 mN and no chemical reaction takes place.

Similar deformation characteristics is observed by the single-point scratching possesses as illustrated in Fig. 3.2. However, dislocations do appear with increased load in this case, as shown in Figs. 3.2b and c. Upon a further increment of the load, the material exhibits its brittle nature, demonstrated by extensive cracking and dislodging, see Fig. 3.2c. No chemical reaction occurs. Grinding introduces significant amorphous phase transformation as seen in Fig. 3.3 a and at the same time initiates two dislocation systems as illustrated in Fig. 3.3b and c. As shown in Fig. 3.3a, grinding introduces significant amorphous phase transformation but at the same time initiates two systems of dislocations, as presented in Figs. 3.3b and c. In this case, the major deformation is still due to the phase change because only two

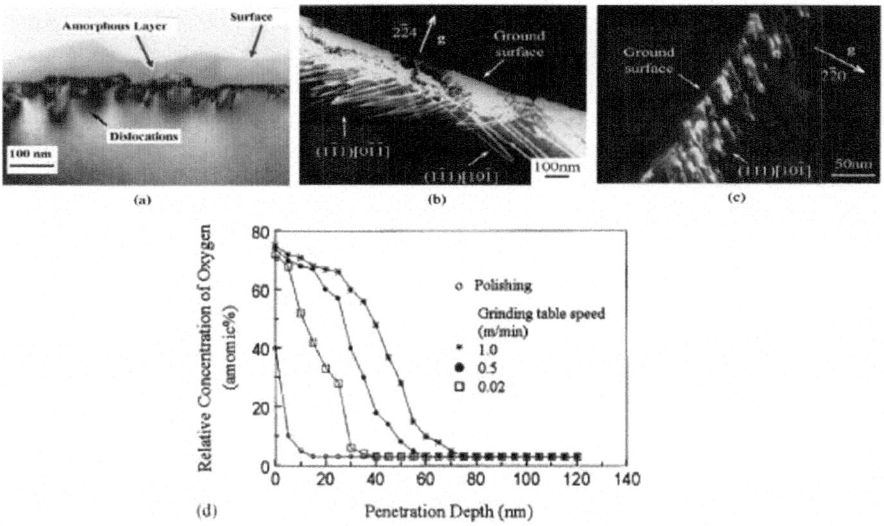

Fig. 3.3 Grinding-induced deformation and oxygen penetration accelerated by grinding and polishing: **a** subsurface deformation with both amorphous transformation and dislocations; **b** dislocations induced by grinding at a table speed of 1 m/min (note that two dislocation systems appear); **c** dislocations induced by grinding at a table speed of 0.02 m/min (note that only one dislocation system appears); **d** variation of depth of oxygen penetration with loading conditions. L. Zhang and I. Zarudi, International Journal of Mechanical Sciences 43, 1985 (2001). With kind permission of Elsevier

Fig. 3.4 A cross-sectional view of the polishing-induced subsurface deformation. Note that dislocations and planar defects do not appear. Amorphous phase transformation with oxidation is the only cause of plastic deformation. L. Zhang and I. Zarudi, International Journal of Mechanical Sciences 43, 1985 (2001). With kind permission of Elsevier

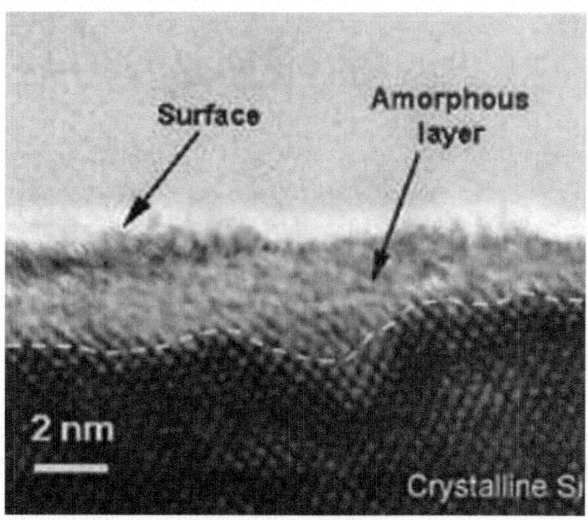

slip systems of dislocations cannot form considerable macroscopic plastic deformation. Analysis of the ground components reveals that grinding accelerates chemical reaction through the application of the water-based coolant. Oxygen penetrates into the deep subsurface of the silicon specimen, whose depth depends on the grinding conditions, as shown in Fig. 3.3d, and forms silicon oxide. Neither planar defects nor the BCC phase of silicon appears.

In summary, increase in temperature accelerates penetration of oxygen into silicon and reduces the critical stress of yielding. Plasticity of silicon in the form of phase transformation (from amorphous to bcc structure) is determined by the loading history. Variation of the depth of oxygen penetration depends (see Fig. 3.3d) on loading conditions.

3.1.2 *In SiC*

SiC powders with various levels of oxygen content were obtained either by acid leaching or heat treatment. These SiC powders were densified by a pressureless sintering technique with boron carbide and carbon additives, which are sintering agents. As known, SiC is an important structural ceramic material because of its high-temperature strength, high hardness, high thermal conductivity, and superior corrosion and thermal shock resistance. These excellent properties have led to a variety of applications of SiC. However, densification of SiC is critically dependent on oxygen content. The purpose of this communication is to present the effect of oxygen impurities on the microstructures, sintering, and mechanical properties of dense SiC ceramics. For this purpose, SiC powders with various levels of oxygen content were obtained by two ways as indicated above, i.e., acid leaching and heat treatment with boron carbide and carbon additives as sintering media. Pressureless sintering technique was used to densify these SiC powders. For experimental details one should consult (Feng et al. 2019). To increase the oxygen content, SiC powders were heat treated at different temperatures. To eliminate oxidation of the SiC powders, the heat treatment temperature should not exceed 800 °C. Consequently, to ensure sufficient sintering activity of the SiC powders, they were heat treated at 500 °C, 600 °C, or 700 °C for 1 h. The oxygen contents of the as-acidleached and as-heat-treated SiC powders are given in Table 3.1.

Table 3.1 Oxygen content (wt%) of as-acid-leached and as-heat-treated SiC powders. D. Feng, Q. Ren, H. Ru, W. Wang, Y. Jiang, S. Ren and C. Zhang, Cer. Inter. 45, 23,984 (2019). With kind permission of Elsevier

Acid/heat	0.2 mol/L HF	Raw powders	500 °C (1h)	600 °C (lh)	700 °C (lh)
Oxygen content	0.55	0.85	1.30	1.65	2.10
Code	SCO0	SCO1	SCO2	SCO3	SCO4

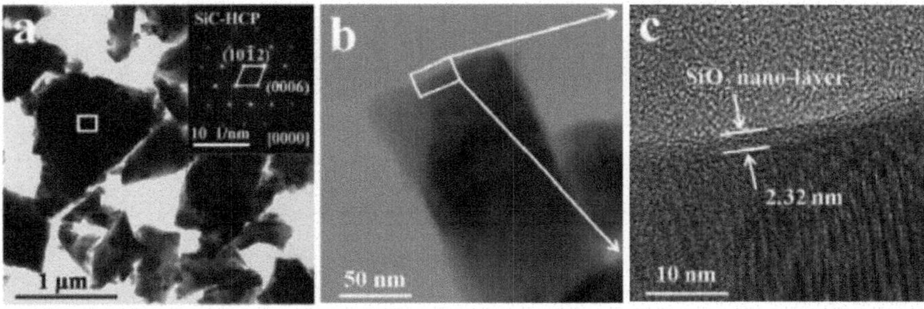

Fig. 3.5 TEM **a**, STEM **b**, and corresponding HRTEM **c** images of as-heattreated SiC powders (SCO4). D. Feng, Q. Ren, H. Ru, W. Wang, Y. Jiang, S. Ren and C. Zhang, Cer. Inter. 45, 23,984 (2019). With kind permission of Elsevier

Images of heat treated SiC powders are illustrated in Fig. 3.5. The relative density and apparent porosity of sintered samples is seen in Fig. 3.6. Note that the relative density decreases with increasing the O content and clearly the apparent porosity increases. SEM images and grain size distributions of the etched surfaces indicate a significant difference in grain sizes between the samples with low and high oxygen content. As the oxygen content increases, abnormal growth and coarsening of the SiC grains occur. The average grain sizes of the samples SCO0–SCO3 are evaluated to be about 4.19, 5.27, 5.74 and 6.30 μm, respectively. The oxygen impurity leads to the formation of gaseous species (CO) which promotes vapor-phase sintering (a non-densifying mechanism), and thereby induces excessive grain growth and coarsening during sintering.

Further, in Fig. 3.6 in the region of low oxygen content, the linear shrinkage ratio of samples hardly changes when the oxygen content increases from 0.55 to 1.30 wt%, but the linear shrinkage ratio decreases significantly in the high oxygen content region (1.65–2.10 wt%). On the other hand, it is clear that the weight loss ratio increases rapidly with an increase in oxygen content (Fig. 3.6). Thus, the oxygen content affects the sintering shrinkage of the samples and reacts with carbon to form volatile species, which leads to the weight loss of the samples. Obviously, the presence of oxygen in the powders is detrimental to the sintering process. This result can be attributed to the gaseous products (CO) formed during high-temperature sintering. The oxygen content of SiC powders should be properly controlled to facilitate the preparation of dense ceramic products.

The effect of oxygen on the mechanical properties are illustrated in Figs. 3.7 and 3.8.

The flexural strength and fracture toughness of SiC samples are shown in Fig. 3.7.

The importance of the oxygen role can be visualized from Fig. 3.8, illustrating the hardness variation with the oxygen content. A significant drop in the hardness value is observed as the oxygen content increases, which indicates that the presence of oxygen impurities in the powders seriously degrades the mechanical properties of the sintered samples. Field emission scanning electron microscopy (FFESEM) is used to investigate the morphology of the sintered fractured surface after flexural

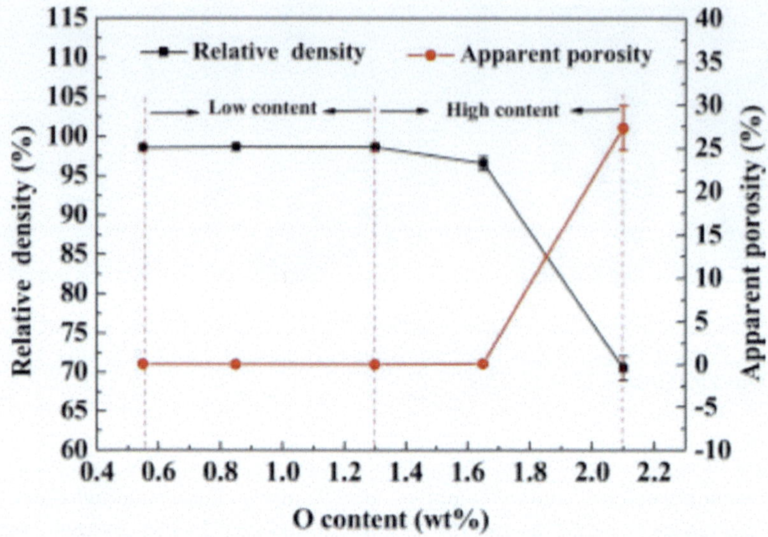

Fig. 3.6 The relative density and apparent porosity of sintered samples. D. Feng, Q. Ren, H. Ru, W. Wang, Y. Jiang, S. Ren and C. Zhang, Cer. Inter. 45, 23,984 (2019). With kind permission of Elsevier

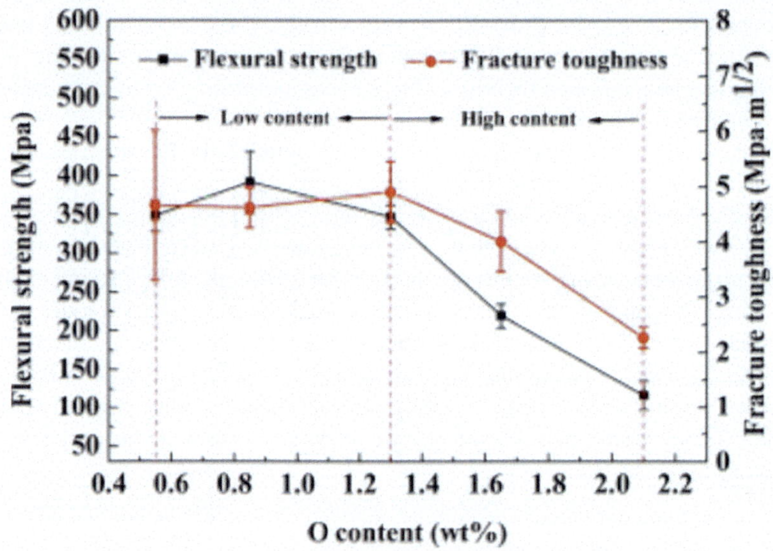

Fig. 3.7 The flexural strength and fracture toughness of samples SCO0–SCO4. D. Feng, Q. Ren, H. Ru, W. Wang, Y. Jiang, S. Ren and C. Zhang, Cer. Inter. 45, 23,984 (2019). With kind permission of Elsevier

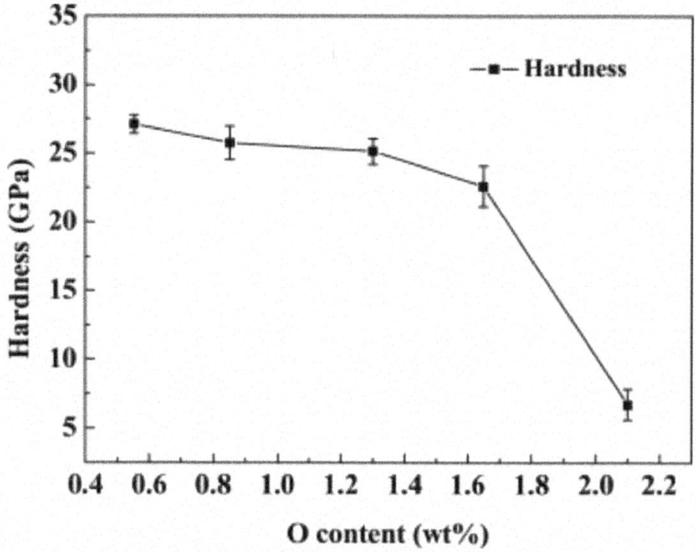

Fig. 3.8 The hardness variation of samples SCO0–SCO4. D. Feng, Q. Ren, H. Ru, W. Wang, Y. Jiang, S. Ren and C. Zhang, Cer. Inter. 45, 23,984 (2019). With kind permission of Elsevier

test and the results are seen in Fig. 3.9. It can be seen from Fig. 3.9 that the fracture characteristic of all samples is mostly transgranular fracture, and relatively smooth and flat regions are observed as a result of crack propagation through the grains.

In summary, the harmful effect of oxygen impurities on the densification process in SiC was considered. The oxygen impurities cover the surface of SiC particles by amorphous silica and the oxygen content of the powders strongly affects the microstructure, sintering and the mechanical properties. Higher oxygen content in the samples accelerates carbon thermal reduction leading to the formation of more gaseous products (CO) and increasing the porosity of the sintered samples, which impedes the densification process. Lower oxygen content samples show smaller average grain sizes and thus better mechanical properties. For good sintering and good mechanical properties, the optimal oxygen content of the powders should be limited to less than 1,30 wt.%.

3.2 The Effect of Moisture

3.2.1 In Si

A significant plastic deformation is related to fatigue, where ~90% of all fractures occur. It is thus a very important deformation, and one might be interested in the effect of various factors. Environmental effects such as moisture in air or the relative

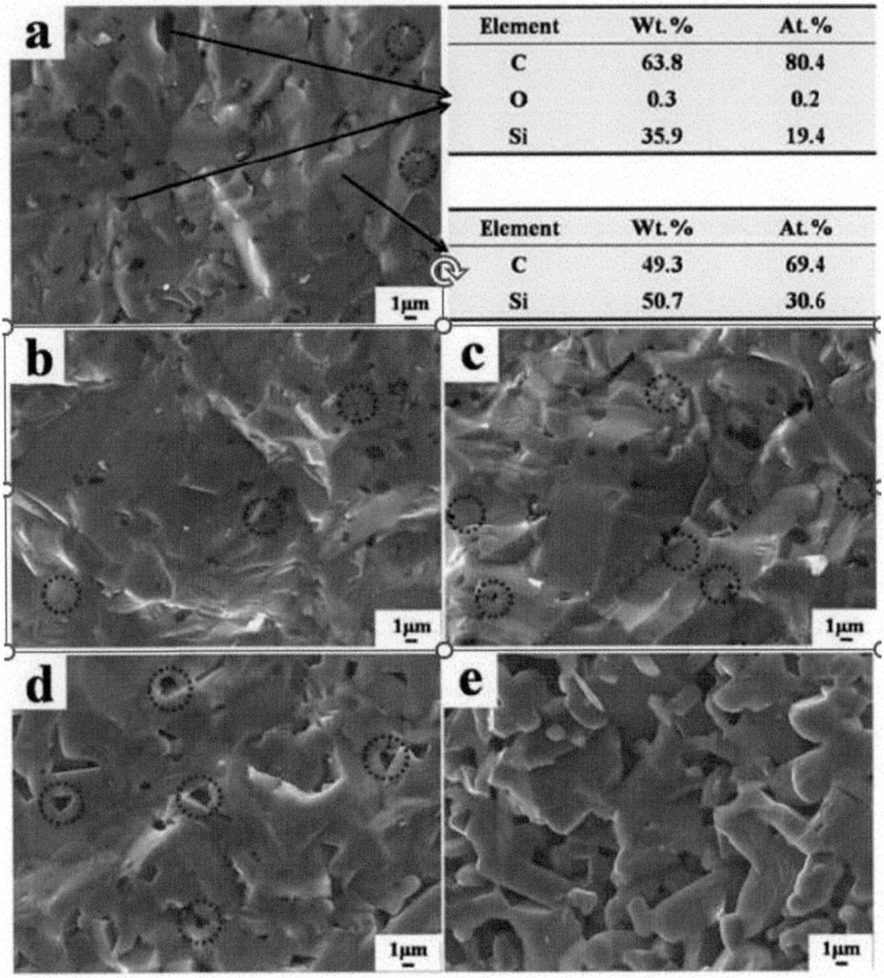

Fig. 3.9 The fracture surface images and EDS analysis of samples: **a–e** represent SCO0 ~ SCO4, respectively. D. Feng, Q. Ren, H. Ru, W. Wang, Y. Jiang, S. Ren and C. Zhang, Cer. Inter. 45, 23,984 (2019). With kind permission of Elsevier

humidity is such a factor to be considered. Understanding the mechanical degradation of a structural material in a micro-electro-mechanical system (MEMS) device is crucial for reliable use.

The susceptibility of a silicon microstructure to environmental conditions is well known. Mechanical properties such as density and elastic modulus vary with temperature. It has been reported that plastic deformation or ductile behavior can be observed in single crystal silicon at low temperature close to room temperature when the representative length of the structure becomes smaller down to the microscale or nanoscale. As long as silicon is protected from the surrounding environment by sealing in a high-vacuum region, silicon shows long-term stability under resonant condition, but fracture (fatigue) is accelerated by the existence of moisture. The effect of the combination of temperature and humidity is significant to the mechanical behavior. The fatigue tests were conducted in an environmental chamber capable of controlling temperature and humidity and is shown in Fig. 3.10. The test consisted of four series at coefficient value (0.25) is almost independent of doping elements. At lower humidity (30% RH), the effect of doping elements (Al, Mg and P) is more pronounced. In undoped condition, the friction coefficient is ~0.8. The friction coefficient decreases to ~0.6 with the addition of 2100 ppm Al and further reduces to 0.45 with Mg (3100 ppm) or P (440 ppm) additions. The reduction in the coefficient of friction values is attributed to chemical kinetics (faster oxidation but controlled hydrolysis) and microstructural (change in the grain boundary structure) factors. The mechanical damage coupled with dopant-controlled hydrolysis reactions play an important role at lower humidity (30% RH), whereas at higher humidity (60% RH), tribochemical reactions (oxidation and hydrolysis) are dominant. As the humidity increases from 30 to 60%, the nature of damage changes from mechanical to tribochemical. Test were performed at two different temperatures and two different humidity conditions. Test conditions are shown in Table 3.2. The structure resonates approximately at the frequency of 49.4 kHz.

Figure 3.11 shows the frequency responses at different temperatures and humidity conditions. As the humidity changed at the same temperature, the resonant frequency and quality factor (Q) were almost the same as illustrated in Fig. 3.11a. However, the resonant frequency decreased as the temperature is increased as seen in Fig. 3.11b. The young modulus for single crystal silicon is given as

$$\frac{E_{Si}(T)}{E_{Si}(20°C)} = 1.019 - 9.385 \times 10^{-4} \quad (3.1)$$

Fatigue properties under different environmental conditions are shown in Fig. 3.12, comparing the effect of humidity. Fatigue life, which is the number of cycles until fracture, increases with decreasing amplitude of applied stress under all environmental conditions. At 25 °C (Fig. 3.12a), the S–N curve is higher at 30%RH than at 85% RH. This result indicates that the fatigue strength of single-crystal silicon was greater under lower humidity condition. It is noted that the fatigue life is more sensitive to the applied stress and the fatigue strength is relatively high at high temperature under low-humidity condition. At a high cycle range over 10^8 cycles, the fatigue strength increases with decreasing humidity at any temperature.

In summary, fatigue tests of single-crystal silicon using a MEMS resonator were performed at different temperatures and humidity conditions. The measurement results demonstrate that humidity has a large effect on fatigue strength and fatigue life, particularly at low temperature. The frequency response was related to the change in Young's modulus. The resonant frequency decreased as the temperature is increased.

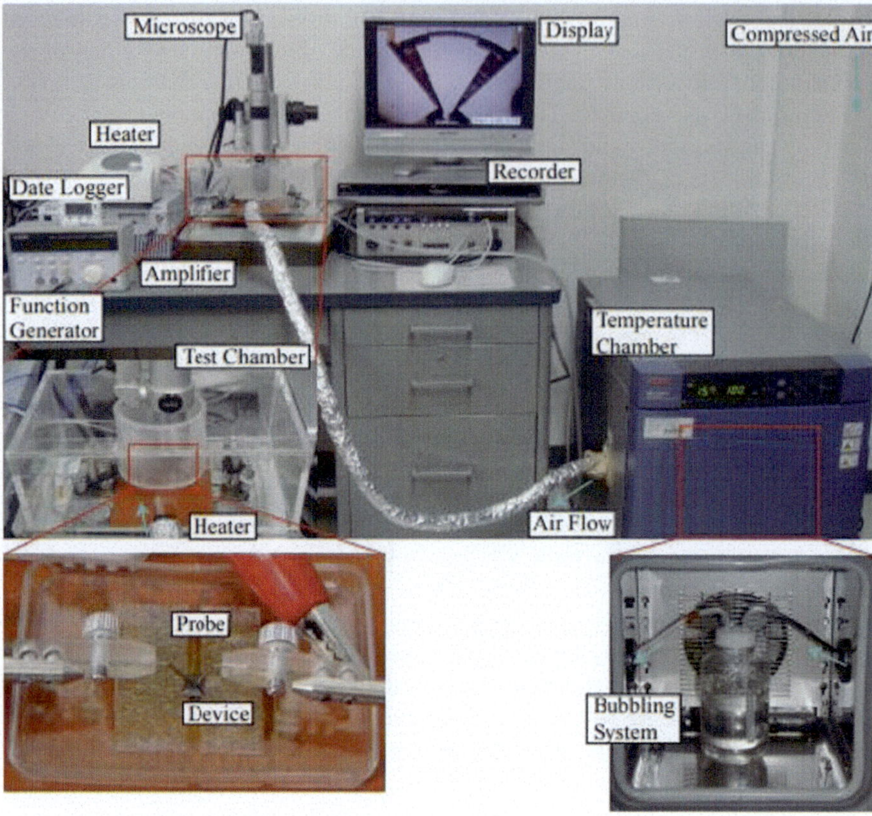

Fig. 3.10 Experimental setup with environmental chamber. T. Ando, M. Shikida and K. Sato, Sensors and Materials, 28, 113 (2016). Open access

Table 3.2 Environmental conditions in fatigue test. T. Ando, M. Shikida and K. Sato, Sensors and Materials, 28, 113 (2016). Open access

Temperature (°C)	25		70	
Relative humidity (RH) (%)	30	85	0.5	15
Absolute humidity (g/m^3)	6.918	19.60	0.9861	29.58

3.2.2 *In SiC*

The effects of relative humidity and doping elements on the friction and wear properties of SiC in unlubricated condition are presented in this section. At higher humidity, the friction coefficient values (~0.25) are low and are independent of doping elements. The effect of small amounts of doping (varying from 400 to 3000 ppm) is more pronounced at lower humidity and the friction coefficient values decrease (as

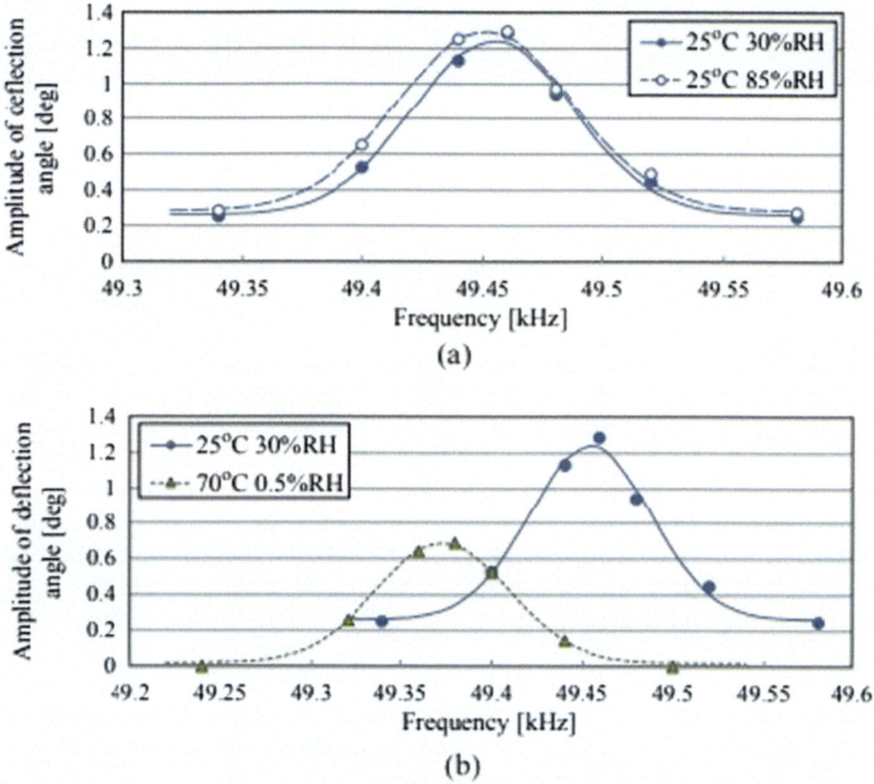

Fig. 3.11 Frequency responses of microresonator in oscillation amplitude at applied voltage of 60 V. T. Ando, M. Shikida and K. Sato, Sensors and Materials, 28, 113 (2016). Open access

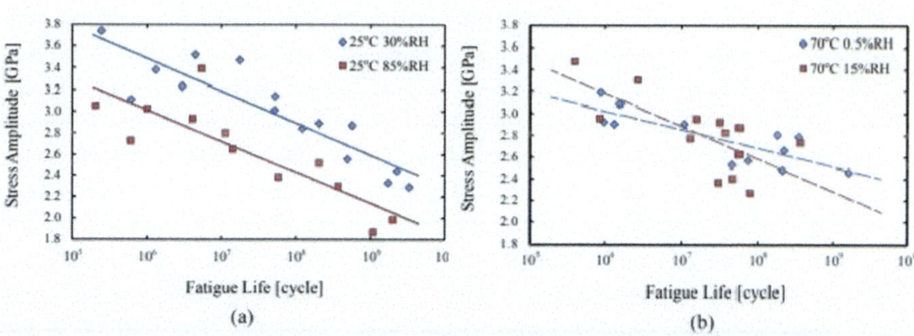

Fig. 3.12 S–N plots of fatigue testing results in various environments at **a** 25 and **b** 70 °C. T. Ando, M. Shikida and K. Sato, Sensors and Materials, 28, 113 (2016). Open access

compared to undoped SiC) with the addition of doping elements such as Al, Mg or P. The specific wear rate is lower at the higher humidity (60% RH) as compared to the lower humidity (i.e., at 30% RH) values. RH stands for relative humidity. As the relative humidity increases from 30 to 60%, the wear mechanism gradually changes from mechanical to tribochemical. The tribo-layer consisting of carbon and hydrated silica progressively smears with relative humidity and becomes instrumental in controlling the friction and wear properties.

As known, structural ceramics are considered as potential tribo-elements, because they exhibit high hardness (essential against friction and wear) and elastic modulus and low thermal expansion with good thermal stability. At RT ceramics are exceptionally stable and do not react with the environment (atmosphere). However, when two ceramics slide against each other, mechanical processes such as indentation, abrasion and attrition produce a large number of tiny debris and the surface area of a material increases by several orders of magnitude. However, The friction and wear properties of dissimilar silicon carbides (α-SiC with 0.37 wt.% of B on doped β-SiC disc with 0.5 Wt.% B) in unlubricated condition are affected by doping as well as humidity. We consider here in particular the effect of humidity. Nevertheless, it should be noted that at higher humidity (60% RH), the friction coefficient value (0.25) is almost independent of doping elements. At lower humidity (30% RH), the effect of doping elements (Al, Mg and P) is more pronounced. In undoped condition, the friction coefficient is ~ 0.8. The friction coefficient decreases to ~ 0.6 with the addition of 2100 ppm Al and further reduces to 0.45 with Mg (3100 ppm) or P (440 ppm) additions. The reduction in the coefficient of friction values is attributed to chemical kinetics (faster oxidation but controlled hydrolysis) and microstructural (change in the grain boundary structure) factors.

A pin-on-disc tester was used to study the dry sliding wear behavior of different SiC tribo-pairs. All tests were conducted in air (relative humidity levels varying from 30 to 60%) for a sliding distance of 130 m maintaining a sliding speed of 36 mm/s (100 rpm) with normal load of 1 N. The average grain size of the ball SiC was ~7 μm. The mechanical properties of the ball material are given in Table 3.3. Only small quantities (up to few thousand ppm) were added to avoid formation of any second phases at the grain boundary. Boron (0.5 wt.%) was added as a sintering aid. The residual doping elements, grain size and mechanical properties of different SiC disc materials are given in Table 3.4. The average grain size of disc samples is ~1 μm. Yaşars used for balls and discs are shown in Fig. 3.13. The variation in the coefficient of friction (COF) as a function of humidity and doping elements is shown in Fig. 3.14. The specific wear rates [= wear volume/(sliding distance × load)], and corresponding humidity values are plotted in Fig. 3.15 and also shown in Table 3.5. The wear rate decreases with relative humidity in the air. At lower humidity, the wear rate was sensitive to doping elements. The wear rate decreases with relative humidity in the air.

In summary, the friction and wear properties of dissimilar silicon carbides (α-SiC with 0.37 wt.% of B on doped β-SiC disc with 0.5 Wt.% B) in unlubricated condition are affected by doping as well as humidity. At higher humidity (60% RH), the friction coefficient values (~0.25) are low and are independent of doping elements.

Table 3.3 Mechanical properties of α-SiC ball material with 3700 ppm of B. V. S. R. Murthy, H. Kobayashi, S. Tsurekawa, N. Tamari, T. Watanabe and K. Kato, Tribology International, 37, 353 (2004). With kind permission of Elsevier

Property	
Density (96 theoretical)	99.3
Hardness (GPa)	23
Fracture toughness (MPa $\sqrt{m}$)	2.2
Elastic modulus (GPa)	410
Flexural strength (MPa)	550

Table 3.4 Mechanical properties of β-SiC disc materials with different dopants. V. S. R. Murthy, H. Kobayashi, S. Tsurekawa, N. Tamari, T. Watanabe and K. Kato, Tribology International, 37, 353 (2004). With kind permission of Elsevier

Material	Dopant (ppm)	Density (% of theoretical)	Average grain size (μm)	Average hardness (GPa)	Average fracture toughness (MPa $\sqrt{m}$)	Average flexural strength[a] (MPa)
SiC	–	98.9	1.15	28	3.3	1050
SiC	Al: 2100	99.1	1.10	24	3.5	760
SiC	Mg: 3100	99.0	2.50	24	3.8	815
SiC	P: 440	99.4	1.12	28	3.6	820

[a]Four-point bed test

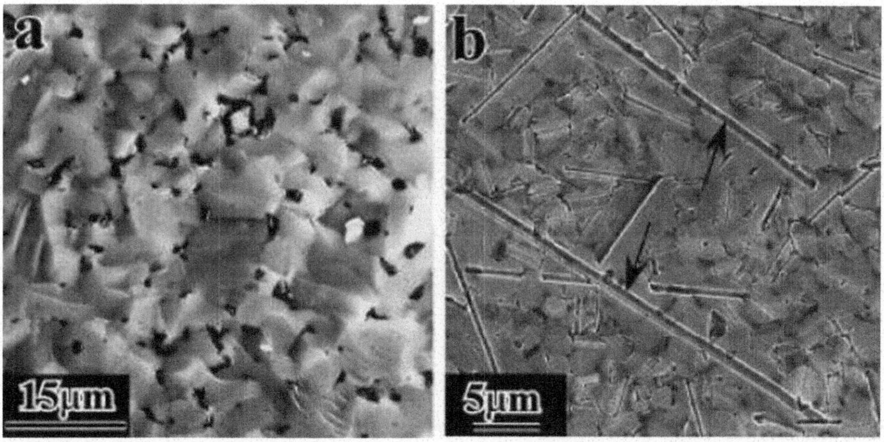

Fig. 3.13 **a** Fracture surface of a α-SiC ball showing the grain size and distribution of graphite as isolated black particles which are smaller than about 1 μm. **b** Etched microstructure of Mg-doped β-SiC disc showing the locations where long needles of graphite (thinner than about 1 μm) appeared. The locations are highlighted by arrows V. S. R. Murthy, H. Kobayashi, S. Tsurekawa, N. Tamari, T. Watanabe and K. Kato, Tribology International, 37, 353 (2004). With kind permission of Elsevier

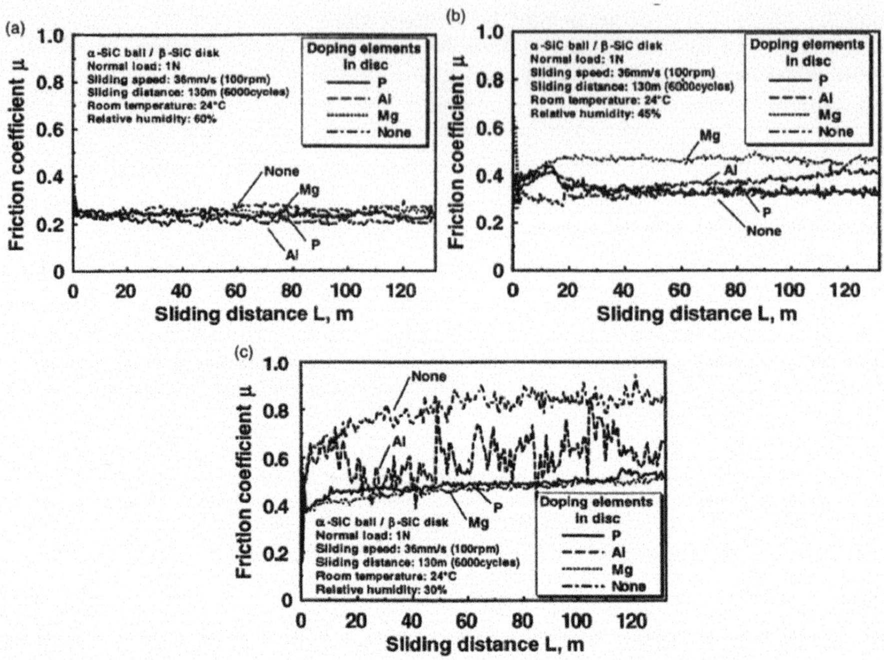

Fig. 3.14 The combined effect of relative humidity and doping elements on the coefficient of friction **a** at 60% RH, **b** 45% RH and **c** 30% RH. V. S. R. Murthy, H. Kobayashi, S. Tsurekawa, N. Tamari, T. Watanabe and K. Kato, Tribology International, 37, 353 (2004). With kind permission of Elsevier

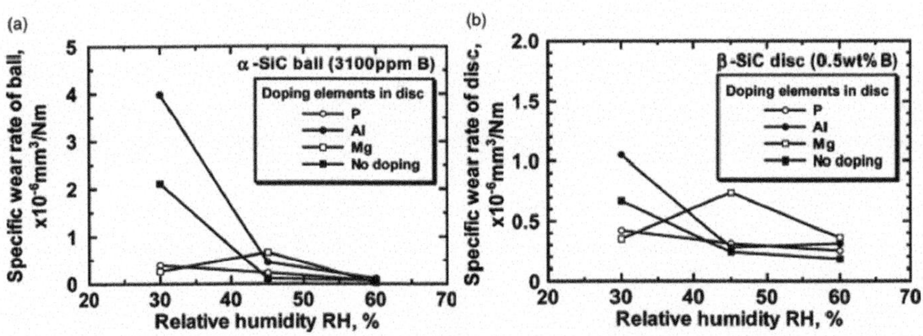

Fig. 3.15 **a** Variation of specific wear rate of α-SiC when slid against different β-SiC disc samples. **b** Specific wear rate of different β-SiC discs (doped as well as undoped) under similar sliding conditions. The wear rate of ball is higher than the disc samples. Note the reduction in the specific wear rate with relative humidity content. V. S. R. Murthy, H. Kobayashi, S. Tsurekawa, N. Tamari, T. Watanabe and K. Kato, Tribology International, 37, 353 (2004). With kind permission of Elsevier

Table 3.5 Specific wear rates of ball and disc in undoped and doped samples with variation of humidity. V. S. R. Murthy, H. Kobayashi, S. Tsurekawa, N. Tamari, T. Watanabe and K. Kato, Tribology International, 37, 353 (2004). With kind permission of Elsevier

Doping elements				
Relative humidity	Al	Mg	P	No dopant
Specific wear rate of ball W_{sb}, $\times 10^{-6}$ $mm^3/N\,m$				
30	3.99	0.31	0.34	2.11
45	0.45	0.66	0.18	0.16
60	0.11	0.02	0.06	0.07
Specific wear rate of disc W_{sd}, $\times 10^{-6}$ $mm^3/N\,m$				
30	1.05	0.37	0.40	0.67
45	0.28	0.74	0.29	0.26
60	0.31	0.34	0.27	0.18

Stress control is a main factor in the operation, performance, and reliability of GaAs devices. Understanding the effect of mechanical stress is important in improvement of the performance reliability of GaAs device design and packaging. GaAs microwave devices which are used for space applications were subjected to mechanical external stress by 4-point bending-by-buckling system. The external mechanical stress was applied on a single GaAs microwave die. To evaluate the value of this stress in device structure and precisely near the channel of the pseudomorphic High Electron Mobility Transistor (pHEMT), simulation based on the Finite Element Method (FEM) has been carried out.

A buckling system shown schematically in Fig. 3.16 was used to quantify the sensitivity of GaAs MMIC dice under external mechanical stress. A uniform in-plane stress (either tensile or compressive) can be applied to the microwave device active surface to change of DC characteristics of the test. DC measurements have been performed with HP 4155 Semiconductor Parameter Analyzer (SPA). Only drain and gate currents have been measured. For both tensile and compressive mode, stress has been increased then decreased and measurements have been made at each step. Material properties and layer thickness are summarized in Table 3.6. The estimated stress (tension and compression) in active layer as a function of loading plate displacement is shown in Fig. 3.17. In the no stress case, seven measurement steps of 5 min have been made whereas with stress seven measurement steps of 24 h have been made. Error estimations are presented in Table 3.7. The change of the threshold voltage and the threshold current of the die pHEMT is plotted in Figs. 3.18 and 3.19. For the present devices, the gate orientation is parallel to x, the loading direction. The threshold voltage V_{TH} increases by + 2.51% under a tensile stress of + 210 MPa and decreases by about − 2.31% at a compressive stress of − 210 MPa.

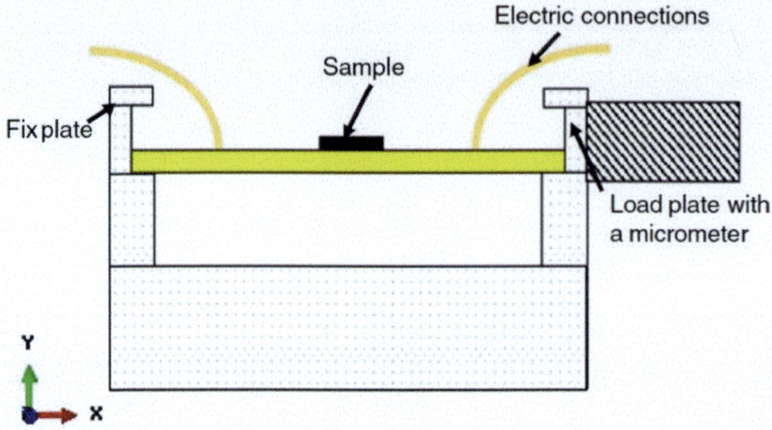

Fig. 3.16 The sketch of the buckling test bench. K. Adokanou, K. Inal, P. Montmitonnet, F. Pressecq, B. Bonnet and J.-L. Muraro, Microelectronics Reliability, 55, 1697 (2015). With kind permission of Elsevier

Table 3.6 Layers material properties and thickness. K. Adokanou, K. Inal, P. Montmitonnet, F. Pressecq, B. Bonnet and J.-L. Muraro, Microelectronics Reliability, 55, 1697 (2015). With kind permission of Elsevier

	Young's modulus (GPa)	Poisson's ratio	Thickness (μm)
PCB	25.8	0.3	400
Die-attach	7.9	0.3	10
GaAs	85.5	0.31	100
SiN_x	150	0.3	0.15

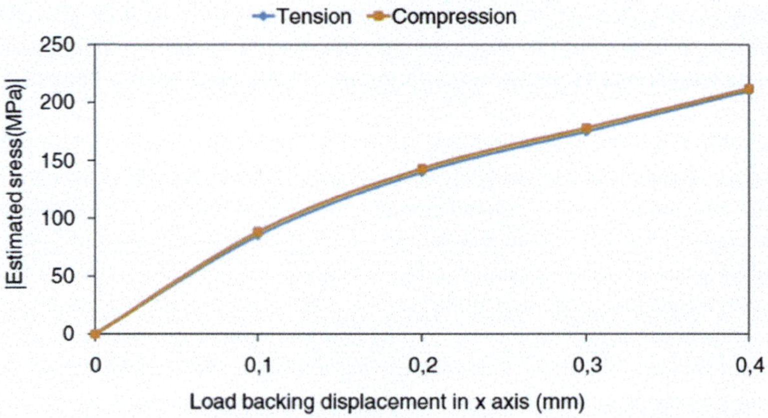

Fig. 3.17 Absolute value of estimated mechanical maximum stress σxx in the active layer as a function of loading plate displacement. K. Adokanou, K. Inal, P. Montmitonnet, F. Pressecq, B. Bonnet and J.-L. Muraro, Microelectronics Reliability, 55, 1697 (2015). With kind permission of Elsevier

Table 3.7 Setup DC characteristic error estimation in %. K. Adokanou, K. Inal, P. Montmitonnet, F. Pressecq, B. Bonnet and J.-L. Muraro, Microelectronics Reliability, 55, 1697 (2015). With kind permission of Elsevier

Characteristics	Voltage	No stress (%)	Applied stress (%)
Threshold voltage	$V_{DS} = 3$ V	0.095	0.115
Threshold current	$V_{DS} = 3$ V/$V_{GS} = 0$ V	0.015	0.096

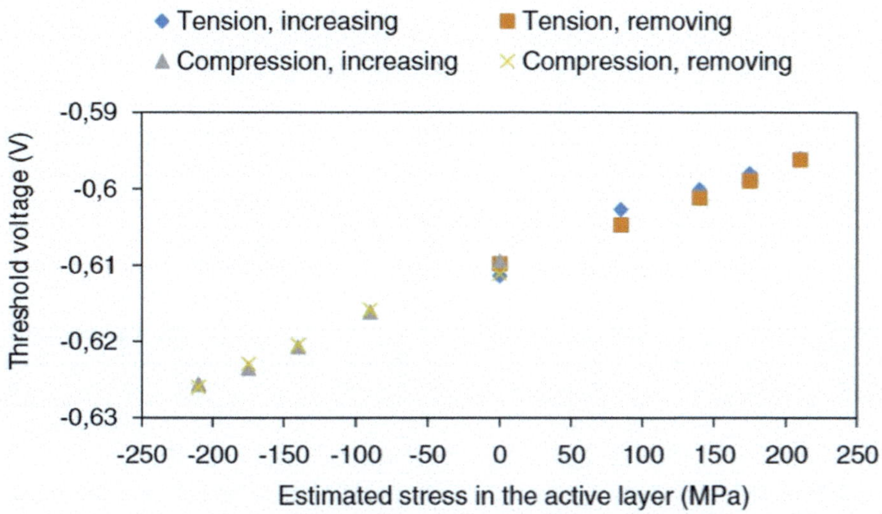

Fig. 3.18 The change of the threshold voltage as a function of the external stress in active layer. K. Adokanou, K. Inal, P. Montmitonnet, F. Pressecq, B. Bonnet and J.-L. Muraro, Microelectronics Reliability, 55, 1697 (2015). With kind permission of Elsevier

Fig. 3.19 Critical stress for dislocation generation from a Knoop indentation plotted against temperature in FZ-Si and CZ-Si. K. Sumino, Met. And Mater. Trans., 30A, 1465 (1999). With kind permission of Springer Nature

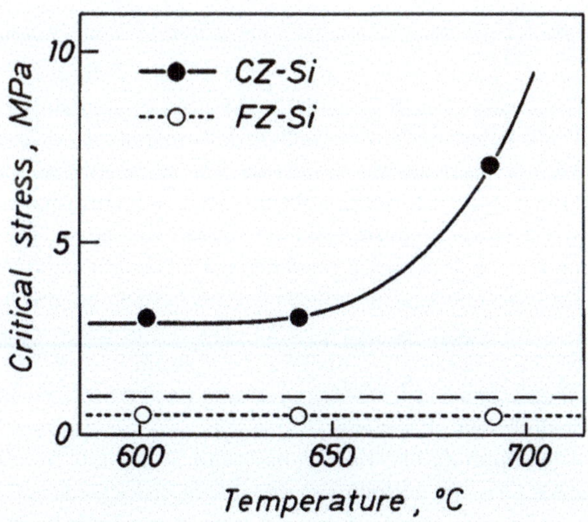

3.3 Tensile Deformation, The Effect of Environment

3.3.1 The Effect of Oxygen

3.3.1.1 In Silicon

Floating zone Si (FZ-Si) and Czochralski-grown Si (CZ-Si) are presented in this section. FZ-Si is a low concentration oxygen ~ 1×10^{16} atoms/cm^3, while the CZ-Si contains about 10^{18} atoms/cm^3. The CZ-Si become saturated at about 1200 °C. Light element impurities such as oxygen in Si have been observed to be effective in suppressing dislocation generation (Kato 2004). Generation of dislocations in CZ-Si has been made by introducing surface flaws such as indentations. Figure 3.19 shows the critical stress for dislocation generation from a Knoop indentation made at room temperature in a FZ-Si and a CZ-Si crystal as a function of the temperature. The development of clusters or precipitates of oxygen along dislocations in the recrystallized dislocated region at a flaw leads to the immobilization of the dislocations. They do not come out of the flaw region until the applied stress exceeds the release stress. Oxygen impurities have been shown to be effective in suppressing the generation of dislocations from such structural irregularities by means of dislocation locking. The velocities of isolated 60 deg and screw types of dislocations in high-purity FZ-Si were measured to be linear with respect to the shear stress in the stress range 1 to 40 MPa and the activation energies to be independent of the shear stress in the temperature range 600 °C to 800 °C.

Figure 3.20 shows the relations between the velocity of a 60° dislocation and the shear stress at various temperatures in the CZ-Si containing oxygen impurities at a concentration of 7.4×10^{17} atoms/cm^3. Data of FZ-Si are also shown in the figure. The velocity of dislocations in CZ-Si decreases more rapidly than in the high purity FZ-Si as the stress is decreased towards the critical stress for the cease of dislocation motion. The same is seen in Fig. 3.21 for CZ-Si at 647 °C. The release stress at 647 °C is seen in Fig. 3.22 with the duration of annealing time.

The behavior of a Si crystal in deformation at a constant strain rate is very sensitive to temperature and the strain rate. The stress increases very rapidly with strain in the beginning of deformation. In this stage the deformation is elastic (See Fig. 23a and c). The strain after the maximum is mostly plastic. The maximum of the stressstrain curve is the upper yield point and the minimum is the lower yield point. Figure 23a is related to temperature. With increase in oxygen concentration the upper yield stress is seen to increase in Fig. 3.24. The deformation relates to 800 °C at a variety of oxygen concentration under a shear strain rate of 1×10^{-4} s^{-1}. All the specimens were annealed at 1300 °C and rapidly cooled to eliminate oxygen precipitation before the deformation test Fig. 3.25.

In summary, the yield behavior of Si crystals free from dislocations is not influenced by oxygen impurities, but that of dislocated crystals is affected significantly by these impurities. All of these characteristics in the yield deformation of Si are shown

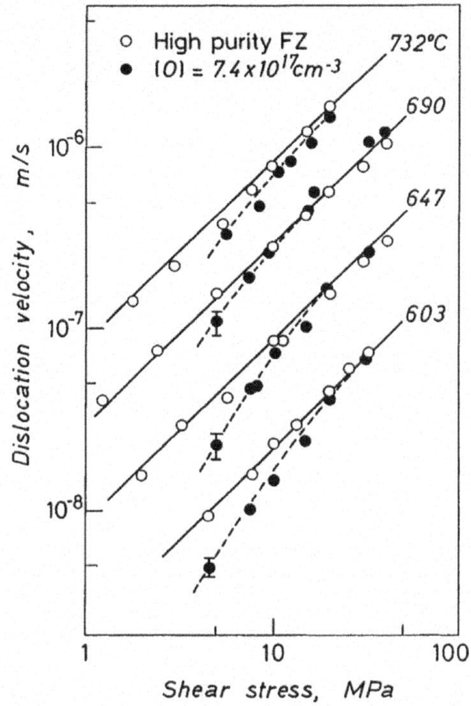

Fig. 3.20 Velocity *vs* shear stress relations at various temperatures of 60 deg dislocations in a CZ-Si crystal containing oxygen impurities at a concentration of 7.4 × atoms/cm^3 to the high purity FZ-Si deformed along the [123] direction under a shear strain 10^{17} atoms/cm^3. Open circles indicate the data for high-purity FZ-Si. K. Sumino, Met. And Mater. Trans., 30A, 1465 (1999). With kind permission of Springer Nature

to be successfully described with a model that considers the dynamic processes of individual dislocations revealed by the direct observations.

3.3.1.2 In SiC

Calculations are performed to study the oxygen doping effects on the mechanical properties of SiC. Both interstitial and substitutional oxygen atoms decrease the shear and tensile modulus and strength of SiC. The substitutional oxygen atoms decrease the shear and tensile modulus and strength more than the interstitial oxygen atoms do. Local distortions around the oxygen doping atoms are found in both of tensile and shear stress–strain curve calculations in the elastic region. The shear strength of SiC decreases linearly with the increase of the oxygen concentration. The tensile strength of SiC also decreases with the oxygen concentration and it drops dramatically at low concentration region. Oxygen impurities may affect the mechanical properties of SiC. Oxygen atoms can stay in both the interstitial site and the substitutional site. The formation energy of oxygen related point defects is calculated. The most stable interstitial oxygen configuration is shown in Fig. 3.26a. In this configuration, the interstitial O_{int} atom is between a C-Si bond and a C-O-Si bridge forms.

Calculations are performed to study the oxygen doping effects on the mechanical properties of SiC. Both interstitial and substitutional oxygen atoms decrease the

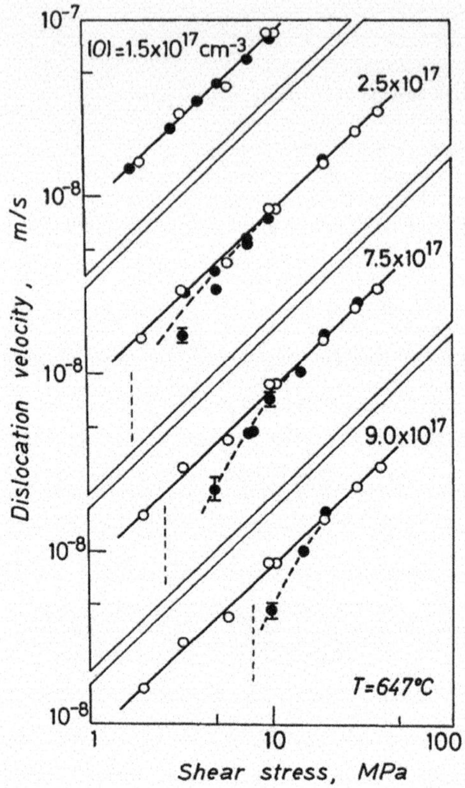

Fig. 3.21 Velocity vs shear stress relations at 647 °C of 60 deg dislocations in CZ-Si crystals containing oxygen impurities at concentrations shown in the figure. Vertical broken lines show the critical stresses for cease of dislocation motion. Open circles indicate the data for high-purity FZ-Si. K. Sumino, Met. And Mater. Trans., 30A, 1465 (1999). With kind permission of Springer Nature

shear and tensile modulus and strength of SiC. The substitutional oxygen atoms decrease. A stress–strain curve of a SiC sample with two oxygen interstitials under shear strain condition is shown in Fig. 3.27. The atomic configurations of the initial point, ultimate strength point, and the broken status are shown in the figure. The shear stress increases with the strain until an inflection point appears, and then the curve increases again. Eventually, the curve reaches the ultimate strength point. After the sample breaks, the stress is released. Since the stress is released, the configuration of the upper part of the sample recovers and there is a deformation layer around the oxygen atoms. It suggests that the sample breaks around the oxygen atoms. In another word, the oxygen doping atom reduces the shear strength. The shear strength and shear modulus of SiC decreases with the increase of the concentration of both interstitial and substitutional oxygen. For the same concentration, the shear modulus and strength of a SiC sample with O_{sub} impurities are lower than that of a sample with O_{int} impurities. The calculated shear modulus and shear strength as functions of interstitial and substitutional oxygen are shown in Fig. 3.28. A strain stress curve for a SiC super cell with two interstitial atoms is shown. The stress is released when sample is broken, and cleavage appears near the oxygen atom. This means that the sample is still brittle since the tensile strength of SiC decreases on

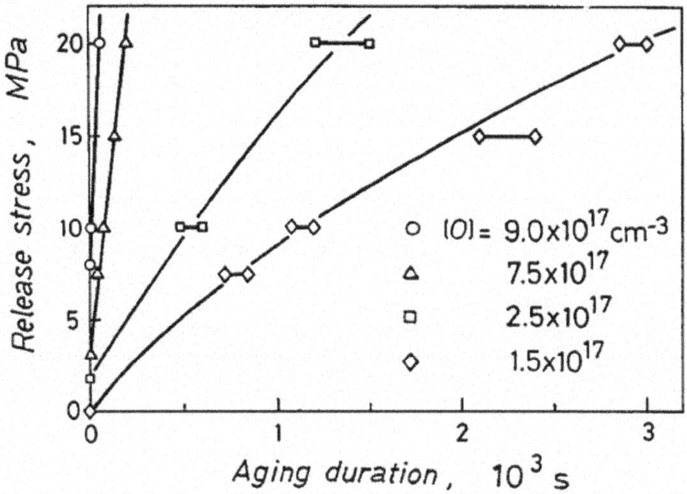

Fig. 3.22 Variation of the release stress at 647 °C for initially fresh 60 deg. Dislocations against the duration of aging at 647 °C in Si crystals containing oxygen impurities at concentrations shown in the figure. K. Sumino, Met. And Mater. Trans. A, 30, 1465 (1999). With kind permission of Springer Nature

doping with oxygen atoms. The mean tensile strength of eight samples is illustrated as tensile stress vs. tensile stain (under tensile strain condition) in Fig. 3.29. The stress increases with the increase in tensile strain linearly, with a small deviation from linearity before the sample is broken. The tensile strength and the modulus as a function of the concentration of interstitial and substitutional oxygen impurities can be seen in Fig. 3.30. The tensile strength and tensile modulus decrease with increase in oxygen concentration, interstitial or substitutional. Local distortion around the doping atoms occurs in the elastic region for both kind of oxygen atoms as seen in Fig. 3.31. The fracture morphologies with two or three oxygen doping atoms under shear and tensile strain conditions are shown in Fig. 3.33. In all the samples the oxygen doping atoms are in the fracture region, which means that the oxygen doping atoms weaken the mechanical properties of SiC. The calculations are validated with data from literature and are listed in Table 3.8 (Fig. 3.32).

In summary, calculations were performed to explore the effects of oxygen doping atoms on the shear and tensile properties of SiC. Both interstitial and substitutional oxygen atoms decrease the tensile and shear modulus and strength of SiC. Substitutional oxygen atoms decrease more the modulus and strength than interstitial oxygen atoms. Local distortion around oxygen atoms in the elastic region was observed. Shear and tensile strength decrease linearly with oxygen concentration.

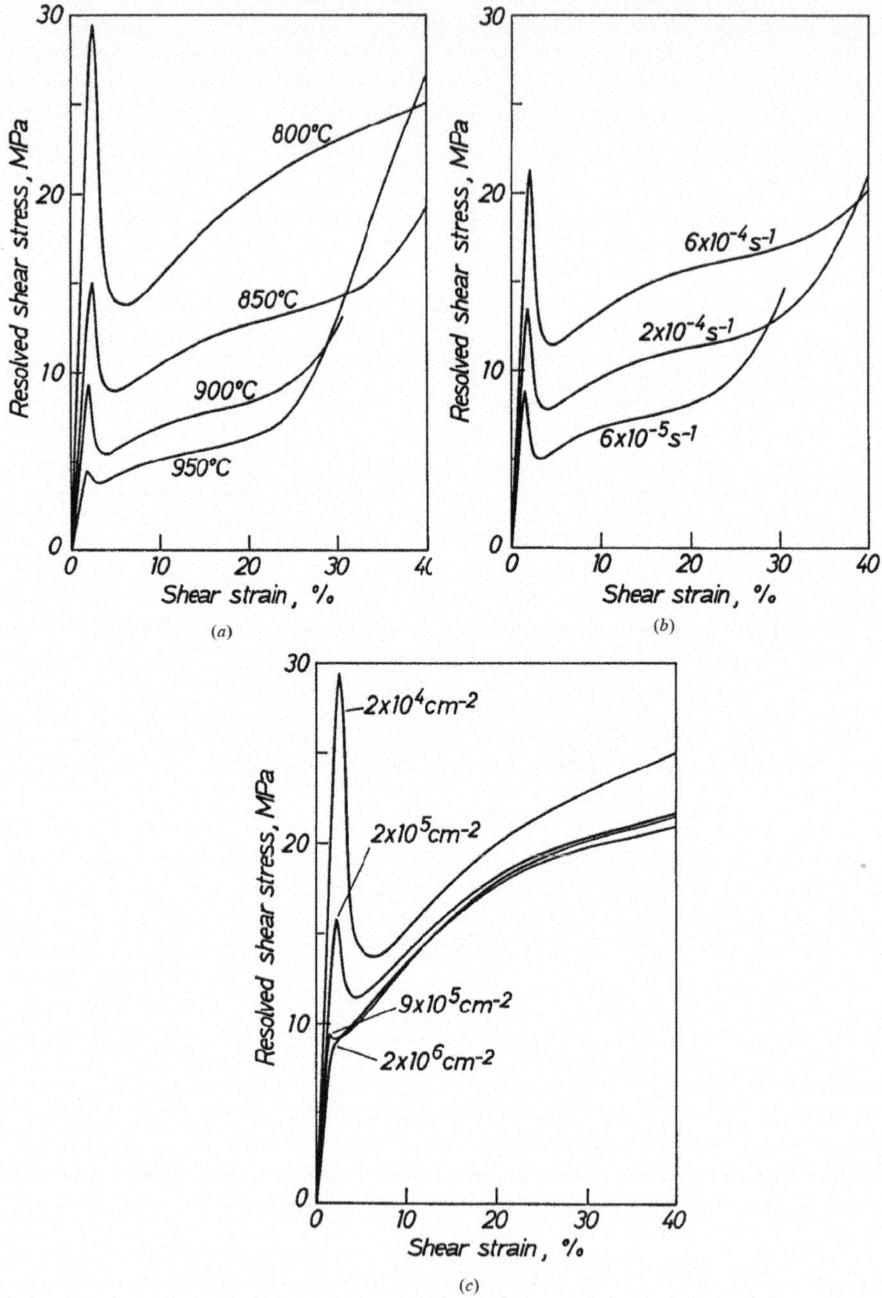

Fig. 3.23 Stress–strain curves of high-purity FZ-Si crystals in tensile deformation along the [123] direction as dependent on **a** temperature, **b** shear strain rate, and **c** initial density of dislocations. K. Sumino, Met. And Mater. Trans., 30A, 1465 (1999). With kind permission of Springer Nature

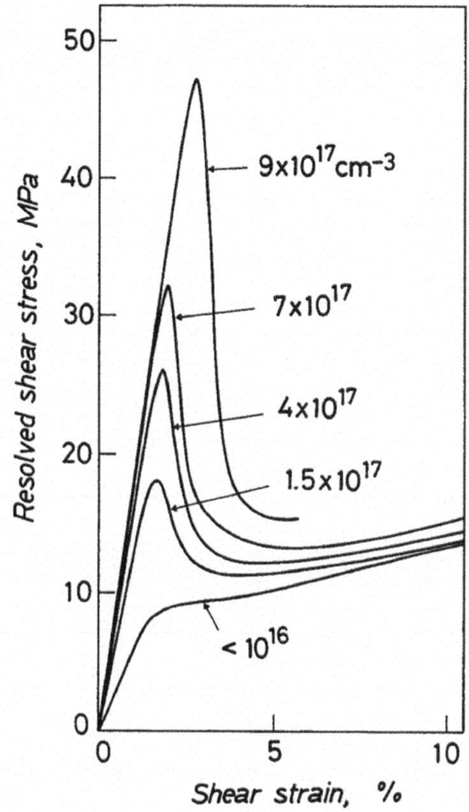

Fig. 3.24 Stress–strain curves of Si crystal dislocated at densities of about 1×10^6 cm^{-2} and containing oxygen at various concentrations shown in the figure deformed at 800 °C under a shear strain of 1×10^{-4} s^{-1}. K. Sumino, Met. And Mater. Trans., 30A, 1465 (1999). With kind permission of Springer Nature

3.3.2 The Effect of Moisture

3.3.2.1 In SiC

Moisture and temperature effects of a known ceramic matrix composite (CMC) based on SiC matrix (Syl-iBN/Bn/SiC) which is composed of Sylramic fibers with an in situ layer of boron nitride (iBN), boron nitride interphase (BN) and silicon carbide matrix are considered in this section. The tests consisted of tensile rupture tests. In such tests the tests are generally performed under tensile stresses less than the yield stress. In the tests laboratory air or pure steam were the environment. The temperatures were 900, 750 and 950 °C. At a given stress level, time to failure decreased with increase of the test temperature and moisture. The primary damage occurred at the BN interface between fiber and matrix, which was relatively more in the steam environment than in the laboratory air.

Carbon and boron nitride are among the most commonly used interphase materials (they are also sintering additives). The preferential use of Syl-iBN/Bn/Si over Si/SiC CMC is due to the oxidation embrittlement in the later system and humid interface

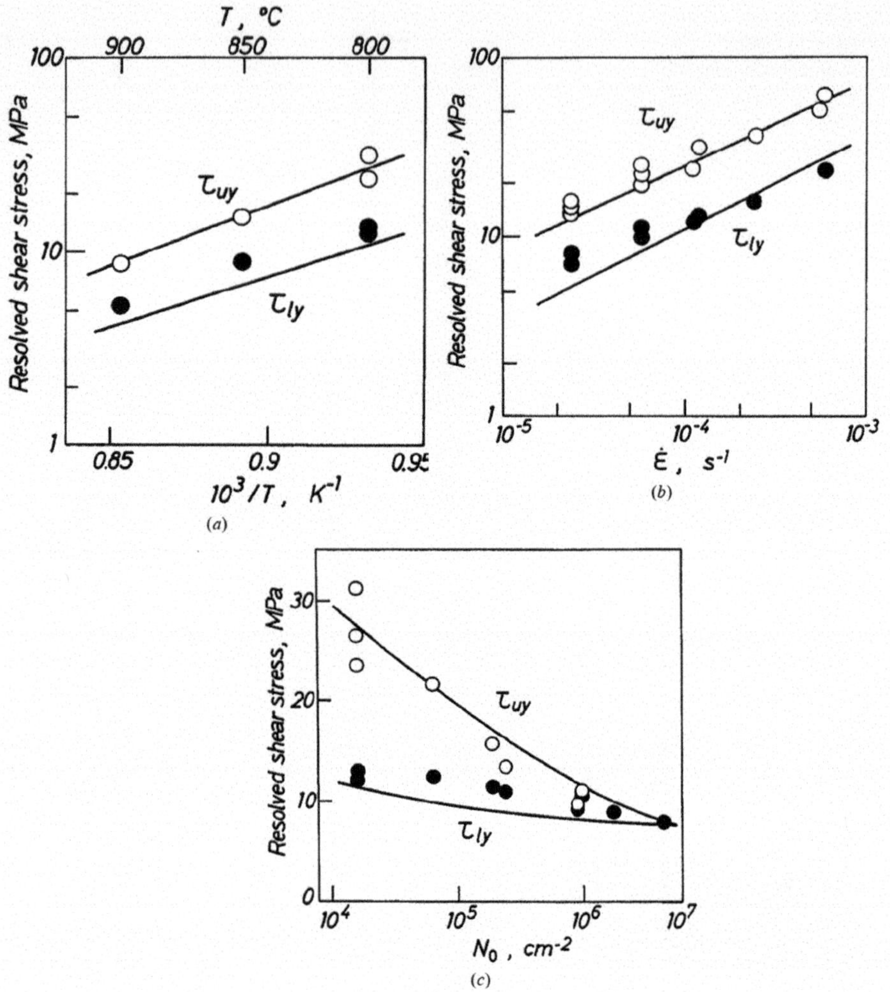

Fig. 3.25 **a** through **c** show the calculated upper and lower yield stresses with solid lines as functions of various parameters. Experimental data are shown by open and full circles. The agreement between calculation and experiment is excellent. K. Sumino, Met. And Mater. Trans., 30A, 1465 (1999). K. Sumino, Met. And Mater. Trans., 30A, 1465 (1999). With kind permission of Springer Nature

penetration at the fiber/matrix interphase. A summary of material properties from the monotonic tests is presented in Table 3.9. The relationship between weight changes and oxidation temperatures for the 3D Hi-Nicalon/SiC specimens after oxidation for 10 h is shown in Fig. 3.34. Weight changes as a function of exposure time for the 3D Hi-Nicalon/SiC composite is shown in Fig. 3.35. Typical load–displacement curves of the composite before and after oxidation are shown in Fig. 3.36. After wet oxidation after the maximum load point, the load drop is much steeper than that after

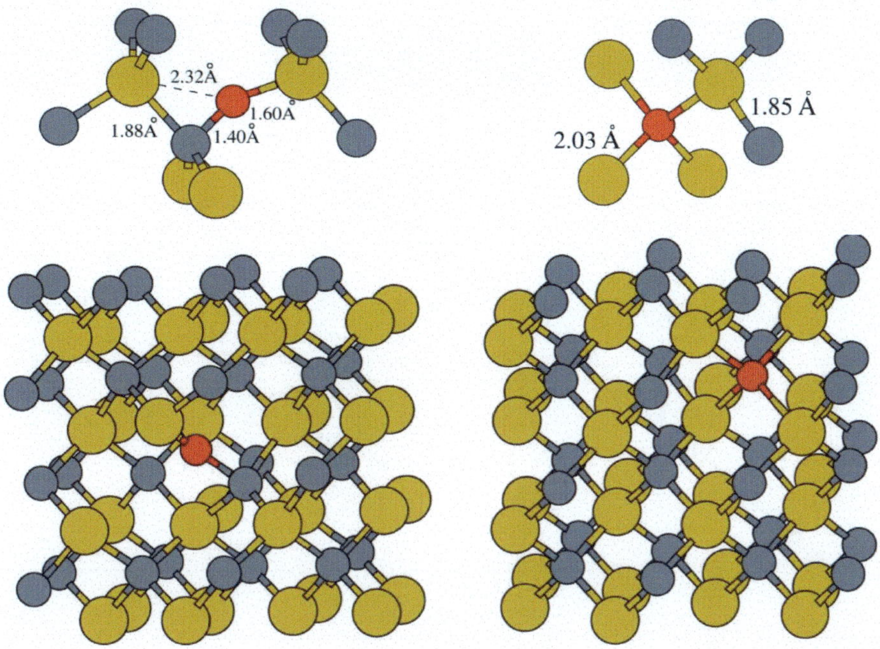

(a) SiC with O_{int} impurity (b) SiC with O_C impurity

Fig. 3.26 SiC super cell with oxygen impurities. **a** with an oxygen interstitial (O_{int}). **b** with an oxygen substitutional at a carbon site (O_C). The yellow, grey, and red balls represent the Si, C, and O atoms respectively. Q. Liu, Y. Li and W. Xiao, Mater. Today Comm. 19, 360 (2019). With kind permission of Elsevier

simulated air oxidation, especially at 1000 °C. Interphase morphology after oxidation at 1000 °C for 10 h in simulated air and wet oxidation is illustrated in Fig. 3.37.

In summary, strength deterioration of the composite is associated mainly with the degradation of the Hi-Nicalon fiber. The difference in fracture behavior is attributed to the change of fiber/matrix interface bonding caused by high temperature oxidation. The higher the test temperature is, the more the failure changes. Oxidation of the 3D Hi-Nicalon/SiC is much enhances in water vapor occurring mainly on the specimen surface. Surface morphologies are strongly affected by oxidation temperature.

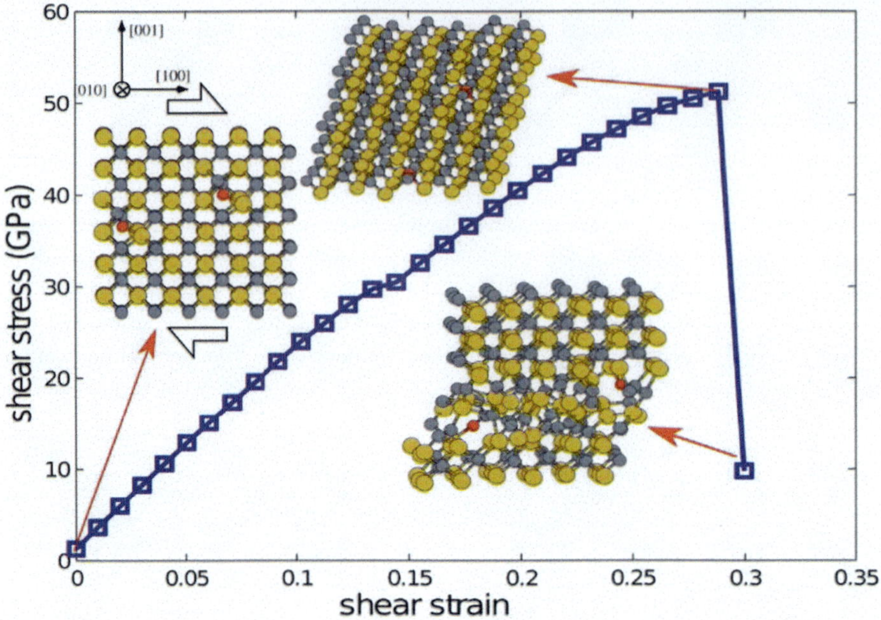

Fig. 3.27 A strain–stress curve of a SiC super cell with oxygen impurities. The configurations of the initial point, ultimate strength point, and the broken status are shown in the figure. The yellow, grey, and red balls represent the Si, C, and O atoms respectively. Q. Liu, Y. Li and W. Xiao, Mater. Today Comm. 19, 360 (2019). With kind permission of Elsevier

3.4 Compressive Deformation the Effect of Environment

3.4.1 The Effect of Oxygen

3.4.1.1 In Silicon

Solar-grade silicon (SoG-Si) contains significantly higher impurity concentrations than the Floating-Zone (FZ) materials (Liu and Y, Li and W. Xiao, 2019). The major electrically inactive impurity in SoG-Si is found to be oxygen. It segregates at dislocations and lock them, preventing their motion and increasing the strength of Si wafers. Oxygen can also precipitate in the bulk material. Inclusions ae actually known to greatly increase the elastic stress in their vicinity making fracture easier in wafer operations such as cutting. The model (finite element) applied (Liu and Y, Li and W. Xiao, 2019) in the polycrystalline Si allows accurate description of the stress–strain behavior of the FZ-Si up to high strains in simple and multiple glide and in various loading conditions (strain rate and temperature). The additional hardening of dissolved oxygen in the bulk is a part of the model.

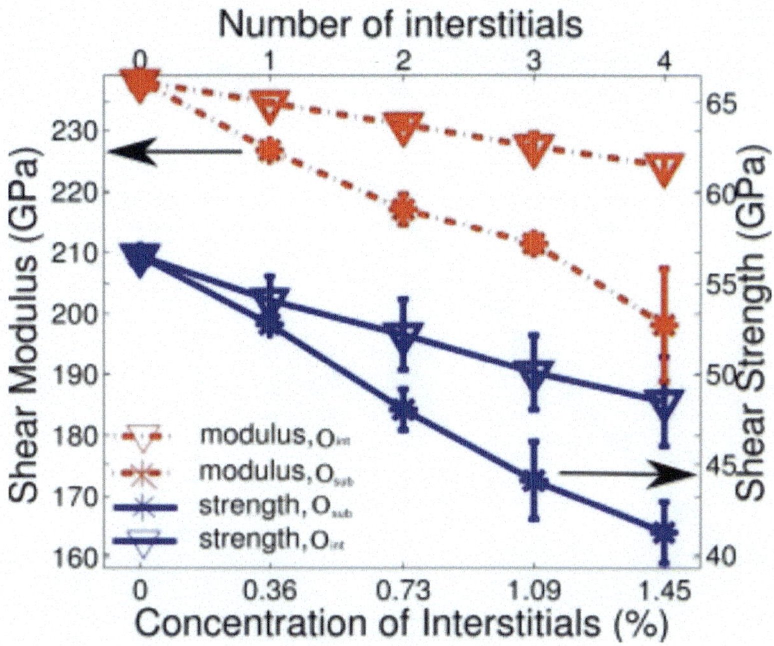

Fig. 3.28 Shear strength and shear module as functions of mass concentration of the interstitial or substitutional oxygen impurities. Q. Liu, Y. Li and W. Xiao, Mater. Today Comm. 19, 360 (2019). With kind permission of Elsevier

The basic assumption of the mode is that plastic deformation of Si crystals proceeds by dislocation glide in their slip system. (For details see (Liu and Y, Li and W. Xiao, 2019)). The ability of hardening stages is shown in Figs. 3.38 and 3.39.

Simulated mechanical behavior of FZ-Si single crystals having the same dislocation densities deformed in the same conditions of temperature and strain rate, but with different oxygen concentration in solution is illustrated in Fig. 3.40.

Dissolved oxygen present in bulk locks dislocations and might prevent their motion, which explains the high strength of dislocated FZ-Si wafers. High values of the stress concentration might lead to crack initiation; high values of oxygen generate high dislocation densities also. Inclusions are frequently found in solar grade silicon and their effect on dislocation generation and on build-up of residual stresses is evaluated by the model (Liu and Y, Li and W. Xiao, 2019).

In summary, dissolved oxygen increases the stress level by limiting dislocation motion. Three-dimensional loading conditions in the bulk lead locally to severe hardening and stress heterogeneities.

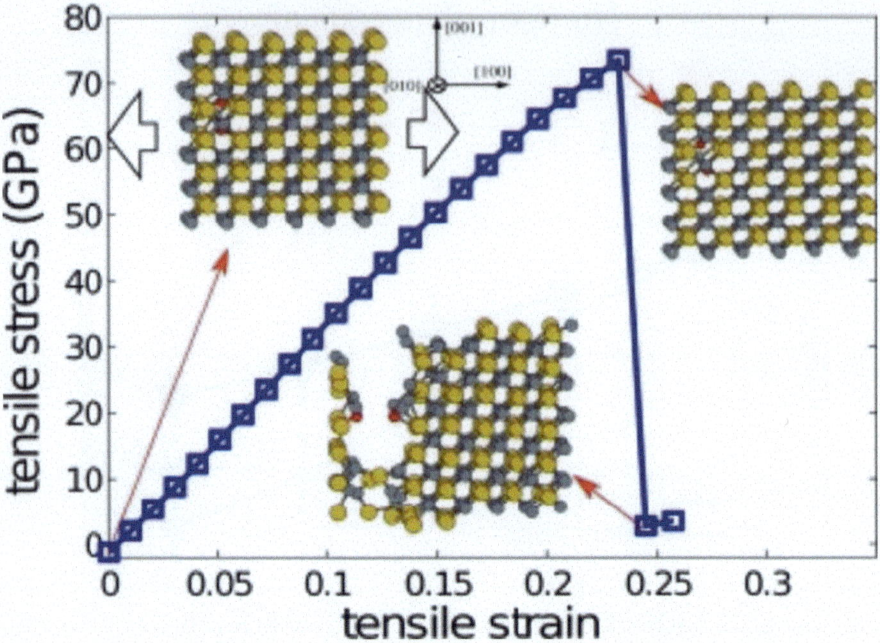

Fig. 3.29 A strain stress curve for a SiC super cell with two interstitial atoms under tensile strain conditions. The configuration of the initial point, ultimate strength point, and the broken status are shown in the figure. The yellow, grey, and red balls represent the Si, C and O atoms, respectively. Q. Liu, Y. Li and W. Xiao, Mater. Today Comm. 19, 360 (2019). With kind permission of Elsevier

3.4.1.2 In SiC

The yielding behavior of powdered beta silicon carbide (SiC) under high pressure and high temperature conditions up to 7.4 GPa and 1400 °C was explored and the results are presented in this section. During compression and heating, data were collected (by x-ray pattern) at each pressure–temperature point, and then, the strains/stresses developed inside the sample under varied pressure–temperature conditions were determined (by assessing the peak width of the x-ray diffraction pattern). No yielding was observed in SiC at cold compression. In contrast, high temperature induces yielding at 1100 °C with a constant loading pressure of ~7.4 GPa. It was found that SiC is the relatively most stable material compared, with other.

Compressed beta-SiC powders of < 10 mm was the starting material in the experiments. Figure 3.41 illustrates the diffraction pattern at ambient. The samples were compressed by step loading from 3 to 60 tons and then heated in steps to 1400 °C at a constant loading of 60 tons. For experimental details see (Mall and Ryba 2013). XRD peaks is used to assess thermal stability in terms of strength at varied high pressure and temperature conditions to evaluate plastic deformation and thus determine the yielding strength. Figure 3.42 illustrates peak broadening as the pressure increases at room temperature. The evolution of peak (111) under high pressure and temperature

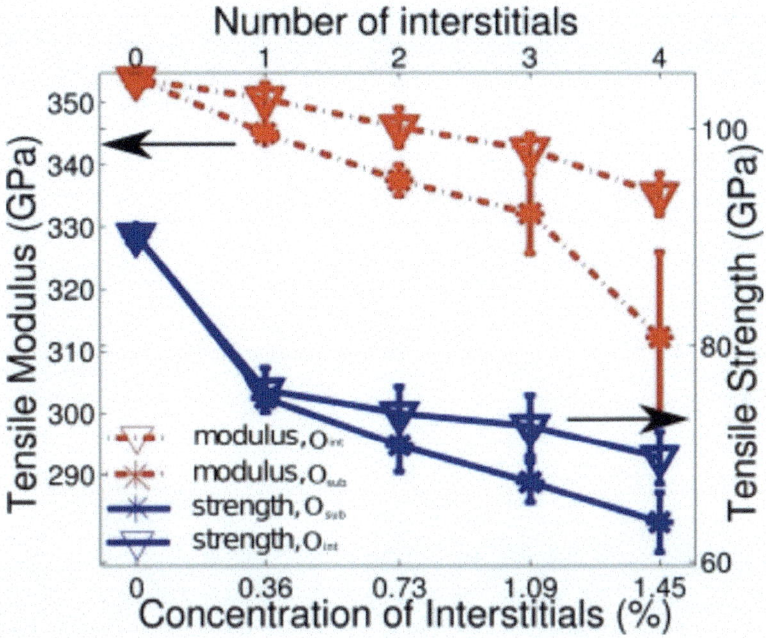

Fig. 3.30 Tensile strength and tensile modulus as functions of the mass concentration of the oxygen impurity. Both interstitial and substitutional configurations are considered. Q. Liu, Y. Li and W. Xiao, Mater. Today Comm. 19, 360 (2019). With kind permission of Elsevier

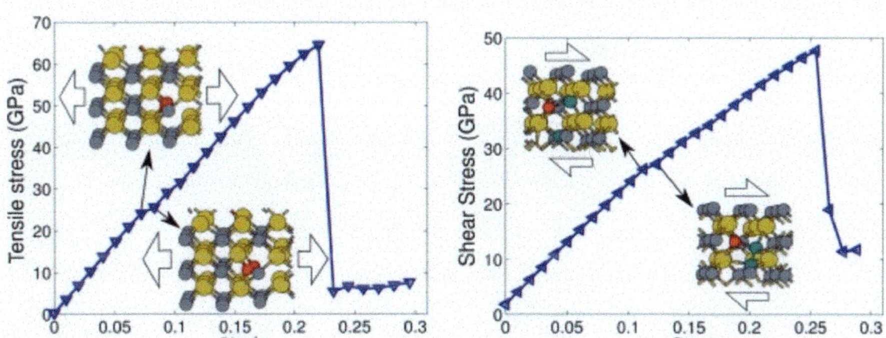

Fig. 3.31 The local distortion happens in the elastic region. **a** A tensile stress strain curve for a SiC super cell with two substitutional atoms. **b** A shear tensile stress strain curve for a SiC super cell with an oxygen interstitial. The two carbon atoms with large displacement are marked in color cyan. Q. Liu, Y. Li and W. Xiao, Mater. Today Comm. 19, 360 (2019). With kind permission of Elsevier

Table 3.8 Equilibrium lattice parameter a_0, bulk modulus B_0, elastic constants C_{11} C_{12} C_{44}, Youngs's modulus E_{100}, tensile strength σ_{100}, shear modulus G_{100}, shear strength τ_{100} of pure SiC. Q. Liu, Y. Li and W. Xiao, Mater. Today Comm. 19, 360 (2019). With kind permission of Elsevier

	This work	Calculations	Experimental
a_0(Å)	4.38	4.32 [39], 4.36 [40]	4.36 [41]
B_0 (GPa)	213	225 [39], 224 [40]	225 [41]
C_{11} (GPa)	387	405 [39], 382 [40]	390 [41]
C_{12} (GPa)	126	135 [39], 145 [40]	142 [41]
C_{44} (GPa)	262	254 [39], 240 [40]	256 [41]
E_{100} (GPa)	354	350 [23], 338 [39]	
σ_{100} (GPa)	90	88 [23], 101 [39]	2.1 [42]
G_{100} (GPa)	238	223 [24], 192 [39]	
τ_{100} (GPa)	57	60 [24], 53 [21]	0.2 [43]

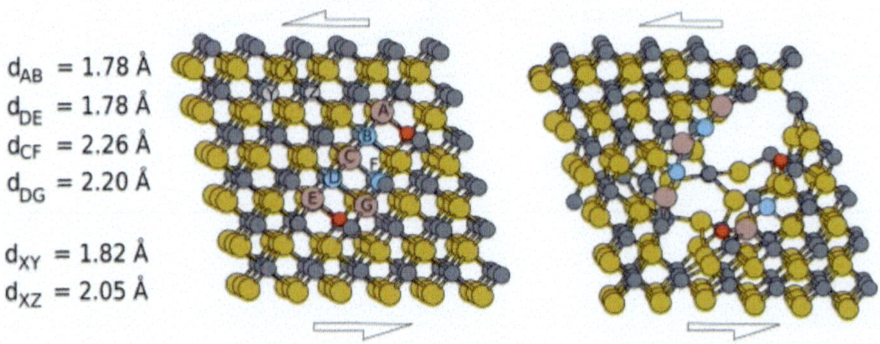

(a) Before fracture (b) After fracture

Fig. 3.32 Configuration of a SiC sample with two interstitial oxygen atoms (atom A to G) on the fracture zte marked with color blue and rosybrown, respectively. The bond length of distorted atoms (atom A to G) on the fracture are marked and not distorted atoms (atom X to Z) are listed. Q. Liu, Y. Li and W. Xiao, Mater. Today Comm. 19, 360 (2019). With kind permission of Elsevier

is seen in this illustration. The apparent strain shown in Fig. 3.43 at given pressures as a function of the ratio Dd^2/d^2 is plotted according to Eq. (3.2) given as d stands for the d spacing of a given lattice plane; Dd_{abs} and Dd_{ins} represent the observed peak width and the portion of peak width caused by instrument resolution, respectively. E and K and L are the differential (deviatoric) strain, constant, and the grain size of SiC.

They were acquired from XRD patterns using a linear fit of the data sets [$d_{abs}{}^2$ / d^2, $d^2(P, T)$], as shown in Fig. 3.43. The apparent stress increases linearly without bending as illustrated in Fig. 3.44.

$$\frac{\Delta d_{abs}^2}{d^2} = \left(\varepsilon^2 + \frac{\Delta d_{ins}^2}{d^2}\right) + \left(\frac{K}{L}\right)^2 d^2(P, T) \tag{3.2}$$

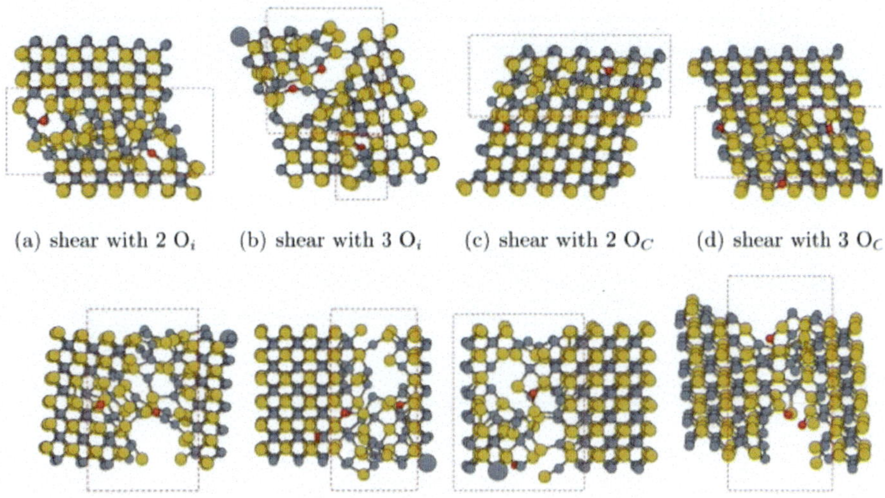

(a) shear with 2 O_i (b) shear with 3 O_i (c) shear with 2 O_C (d) shear with 3 O_C

(e) tensile with 2 O_i (f) tensile with 3 O_i (g) tensile with 2 O_C (h) tensile with 3 O_C

Fig. 3.33 Fracture morphology of SiC samples with 2 or 3 doping atoms under shear or tensile strain conditions. O_i represents an interstitial atom and O_C represents an oxygen atom at the carbon lattice site. Q. Liu, Y. Li and W. Xiao, Mater. Today Comm. 19, 360 (2019). With kind permission of Elsevier

Table 3.9 Test data summary. S. Mall and J. L. Ryba, Composites Science and Technology, 68, 274 (2008). With kind permission of Elsevier

Test conditions	Young's modulus (GPa)	Ultimate strength (MPa)	Normalized ultimate strength[a]	Failure strain (mm/mm)	Source
RT	235	415	1	0.44	Present study
400°C	231	334	0.81	0.36	Present study
750°C	232	327	0.79	0.38	Present
950°C	225	305	0.74	0.41	Present study
550°C	233	347	0.84	0.39	[15]
750°C	235	335	0.81	0.36	[15]

[a]Normalize with the ultimate tensile strength at room temperature

No phase transition was occurring in the SiC sample within the pressuretemperature range investigated. At ~60 GPa and 1700 K starts to decompose to carbon and silicon, indicating the high thermal stability of SiC.

In summary no yielding was observed at cold compression up to 7.4 GPa. However, it occurs at 77.4 GPa and 1100 °C with a yielding stress of 12.5 GPa. Overall, SiC has an outstanding thermal mechanical stability making it suitable for application in high working pressure and temperature environment.

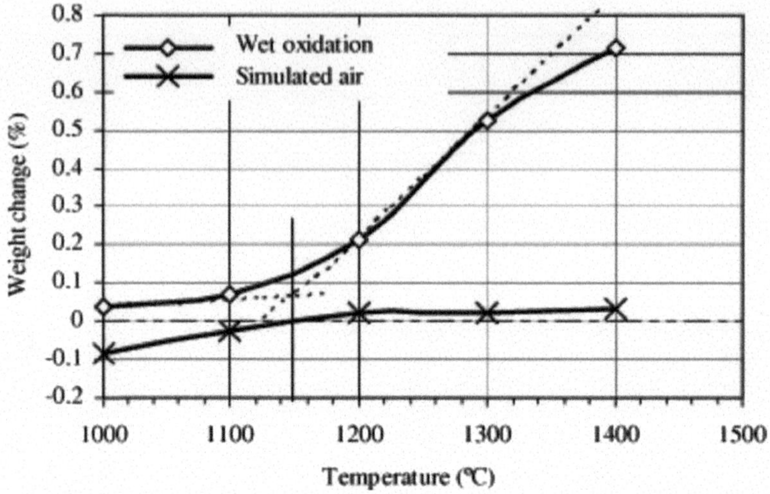

Fig. 3.34 Relationship between weight changes and temperature of the 3D HiNicalon/SiC specimens after oxidation in different oxidizing environments for 10 h. S. Mall and J. L. Ryba, Composites Science and Technology, 68, 274 (22,013). With kind permission of Elsevier

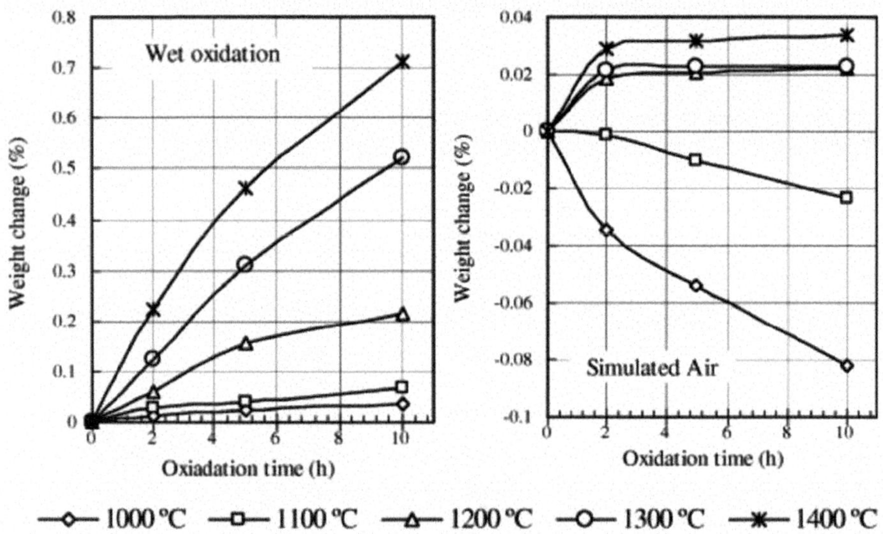

Fig. 3.35 Weight changes as a function of exposure time for the 3D Hi-Nicalon/SiC composite. S. Mall and J. L. Ryba, Composites Science and Technology, 68, 274 (22,013). With kind permission of Elsevier

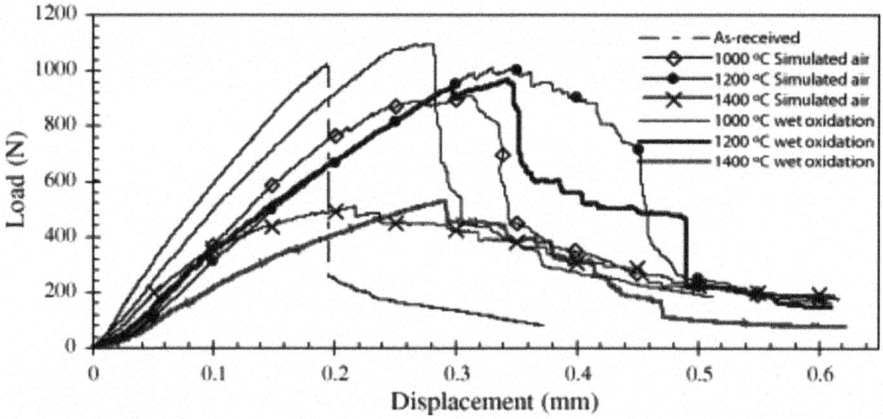

Fig. 3.36 Typical load–displacement curves in flexural strength tests of the 3D HiNicalon/SiC composite after oxidation. S. Mall and J. L. Ryba, Composites Science and Technology, 68, 274 (22,013). With kind permission of Elsevier

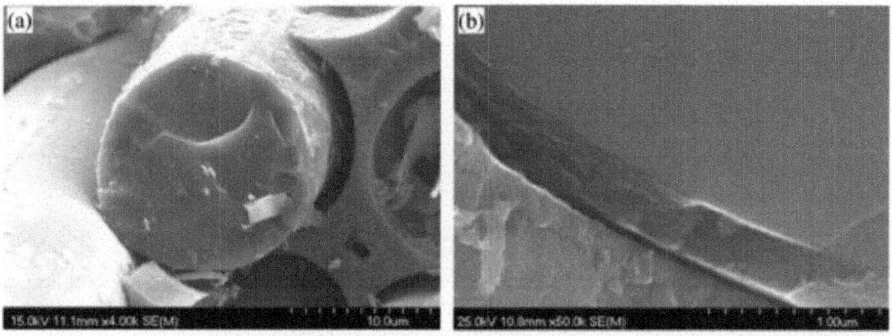

Fig. 3.37 Interphase morphology of the 3D Hi-Nicalon/SiC composite after oxidation at 1000 °C for 10 h in **a** simulated air, **b** wet oxidation. S. Mall and J. L. Ryba, Composites Science and Technology, 68, 274 (22,013). With kind permission of Elsevier

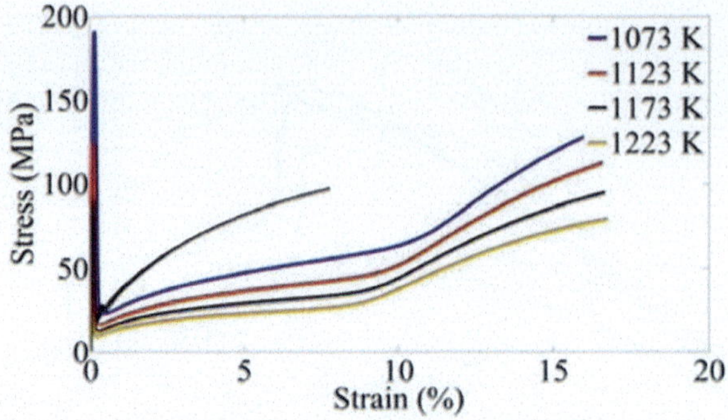

Fig. 3.38 Simulated stress–strain behavior of a FZ-Si single crystal deformed in single glide at different temperatures. The case of double slip is depicted for T = 1173 K (MS curve). J. Cochard, S. Gouttebroze, S. Dumoulin, M. M'Hamdi and Z. L. Zhang, 24th European Photovoltaic Solar Energy Conference, 21–25 September 2009, Hamburg, Germany

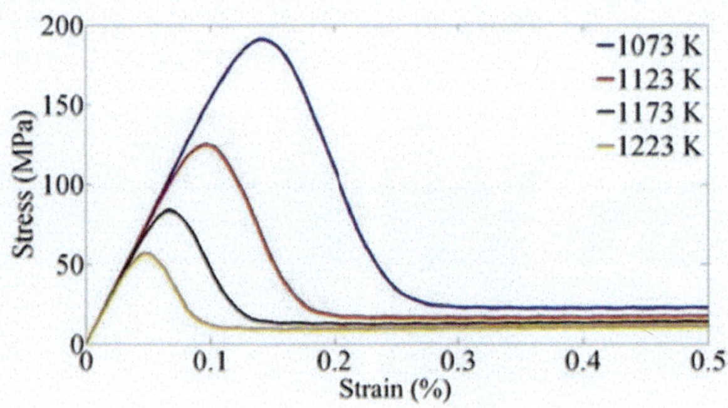

Fig. 3.39 Close-up of Fig. 3.38 at small strains: influence of temperature on the yield region. J. Cochard, S. Gouttebroze, S. Dumoulin, M. M'Hamdi and Z. L. Zhang, 24th European Photovoltaic Solar Energy Conference, 21–25 September 2009, Hamburg, Germany

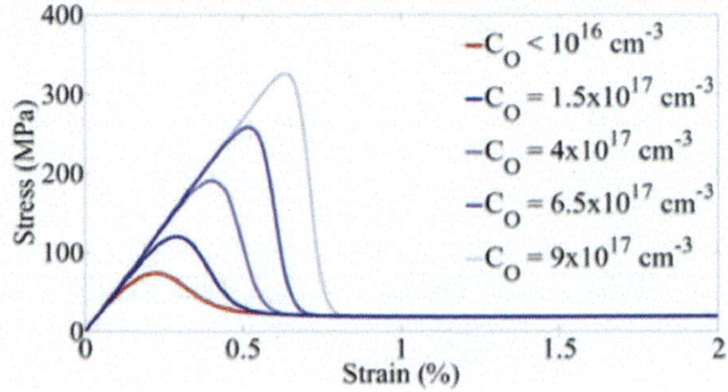

Fig. 3.40 Simulated yield region of a CZ-Si single crystal deformed in single glide for different bulk dissolved oxygen concentrations. J. Cochard, S. Gouttebroze, S. Dumoulin, M. M'Hamdi and Z. L. Zhang, 24th European Photovoltaic Solar Energy Conference, 21–25 September 2009, Hamburg, Germany

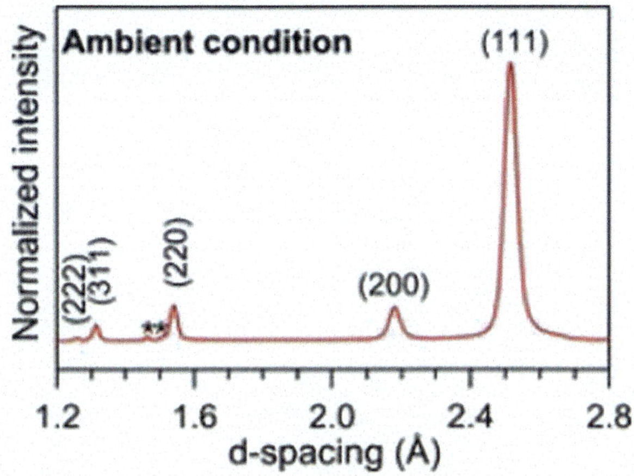

Fig. 3.41 X-ray diffraction pattern of SiC under ambient conditions. The stars mark the fluorescence peaks of lead. Y. Wang, AIP Advances 10, 065,209 (2020). Open access

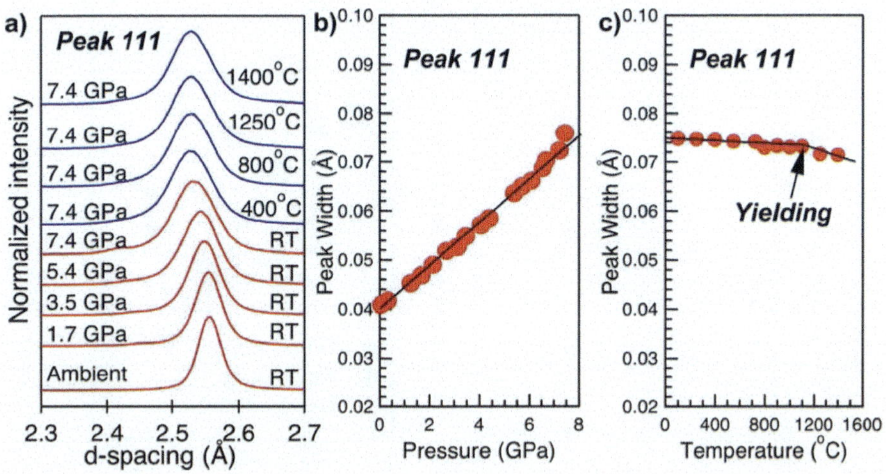

Fig. 3.42 The evolution of peak (111) of SiC under high pressure and temperature conditions. The panel **a** shows the peak (111) collected during compression at room temperature (red colored) and then during heating at 7.4 GPa (blue colored). The peaks become broader as pressure increases, while during heating at a fixed pressure, the patterns still remain broad. The panels **b** and **c** show the variations in the full width at the half maximum of peak (111) vs pressure and temperature, respectively. The peak width increases linearly and gradually up to 1000 °C, after which the decrease in peak width becomes slightly more rapid. Therefore, the heating induces softening of the sample, ad at 1000 °C, the sample yields. Y. Wang, AIP Advances 10, 065,209 (2020). Open access

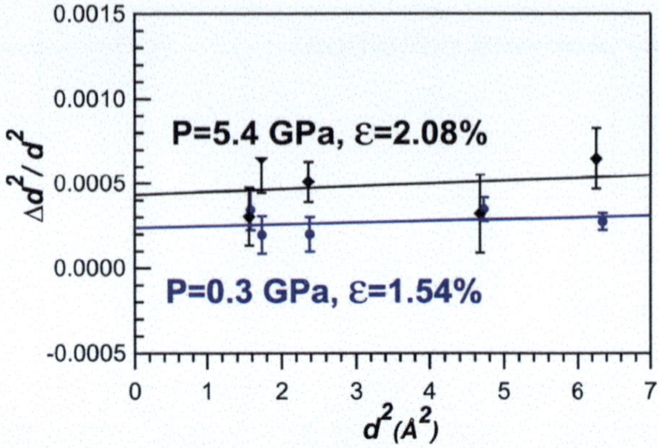

Fig. 3.43 Example of the determination of the apparent strain in the sample at given pressures by using a plot of d^2/d^2 against d^2 according to Eq. (3.2). Y. Wang, AIP Advances 10, 065,209 (2020). Open access

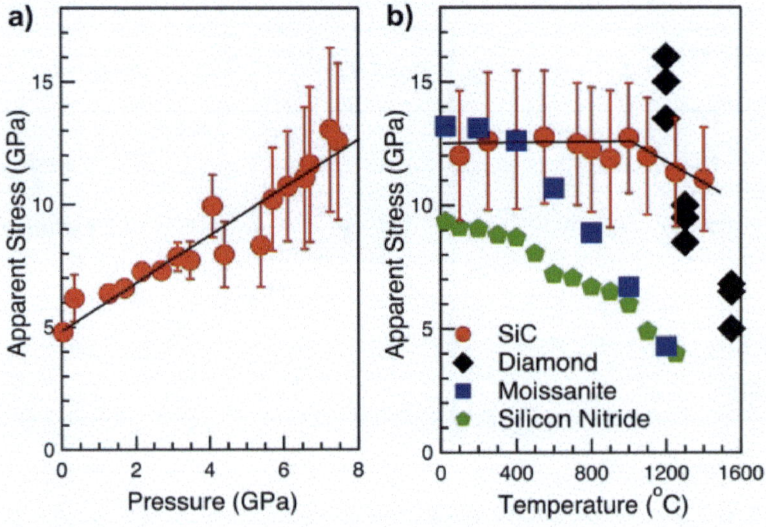

Fig. 3.44 The scheme of the apparent stress as a function of pressure–temperature. The panel **a** demonstrates that the stresses developed in the sample under compression increase linearly with pressure without any yielding. The panel **b** shows that thin the stress in the sample releases linearly and slowly up to 1100 °C, after which the stress inside the sample abruptly drops, forming a yielding point at that temperature with an yielding strength of ~12.5 GPa. For comparison, the variation in diamond, moissanite and silicon nitride during heating at constant pressures are also included in the right panel. Y. Wang, AIP Advances 10, 065,209 (2020). Open access

References

K. Adokanou, K. Inal, P. Montmitonnet, F. Pressecq, B. Bonnet and J.-L. Muraro. Microelectron. Reliab. **55**, 1697 (2015).

T. Ando, M. Shikida, K. Sato. Sens. Mater. **28**, 113 (2016)

J. Cochard, S. Gouttebroze, S. Dumoulin, M. M'Hamdi, Z.L. Zhang, in *24th European Photovoltaic Solar Energy Conference*, 21–25 Sept. 2009 (Hamburg, Germany, 2009)

D. Feng, Q. Ren, H. Ru, W. Wang, Y. Jiang, S. Ren, C. Zhang, Cer. Inter. 23984 (2019)

Q. Liu, Y, Li, W. Xiao. Mater. Today Comm. **19**, 360 (2019)

S. Mall, J.L. Ryba. Compos. Sci. Technol. **68**, 274 (2013)

V.S.R. Murthy, H. Kobayashi, S. Tsurekawa, N. Tamari, T. Watanabe, K. Kato. Tribol. Int. **37**, 353 (2004)

K. Sumino. Met. Mater. Trans. **30**, 1465 (1999)

Y. Wang. AIP Adv. **10**, 065209 (2020)

L. Zhang, I. Zarudi. Int. J. Mech. Sci. **43**, 1985 (2001)

Chapter 4
Hardness and Tensile Properties. The Effect of Environment

4.1 The Effect of Oxygen

4.1.1 In Ge0.5Si0.5

The lack of information on the effect of oxygen environment in pure nanostructure Ge and since the similarity of Ge and Si, it was felt that it might be of interest to present the subject in a $Ge_{0.5}Si_{0.5}$ alloy nanowire. First the mechanical property of this alloy is presented in order to familiarize with this material which will follow with the effect of oxygen.

The highest fracture strength of the NW is for the [111] oriented crystal compared to other orientations, but the [110] orientation exhibits the highest fracture toughness. Tensile loading in the range of 100–600 K is shown in Fig. 4.1a. Following the linear part, further increase in strain induces non-linear increase in stress up to its peak value and then decreases abruptly with only a small increase in strain. The variator of the cross-sectional area with strain-representing the size effect at 300 K is seen in Fig. 4.1b. Further, the variation of the UTS with temperature of the $Ge_{0.5}Si_{0.5}$ alloy nanowire and its Young's modulus are seen in Fig. 4.1i and d for the sizes shown in Fig. 4.1b. Clearly, they decrease with increasing temperature.

The effect of the orientation of the $Si_{0.5}Ge_{0.5}$ alloy on the mechanical properties is illustrated in Fig. 4.2. Further, the presence of defects has a significant effect on the mechanical properties among them vacancies. For example, 2% of randomly distributed vacancies can deteriorate the fracture strength by ~ 40% in graphene. Randomly distributed vacancies introduced by removal of Si or Ge atoms affects the strength. The uniaxial strength of the $Si_{0.5}Ge_{0.5}$ as a function of Si and Ge vacancy concentrations is illustrated in Fig. 4.3.

The failure behavior of $Si_{0.5}Ge_{0.5}$ occurs at different strain values for 100 K and 600 K as shown in Fig. 4.3. The process of fracture depends on the mechanism of crack propagation. In ductile material, plastic deformation occurs generally ahead of

J. Pelleg, *The Influence of Environment on Mechanical Characteristics of Key Semiconductor Components*, https://doi.org/10.1007/978-3-031-34530-2_4

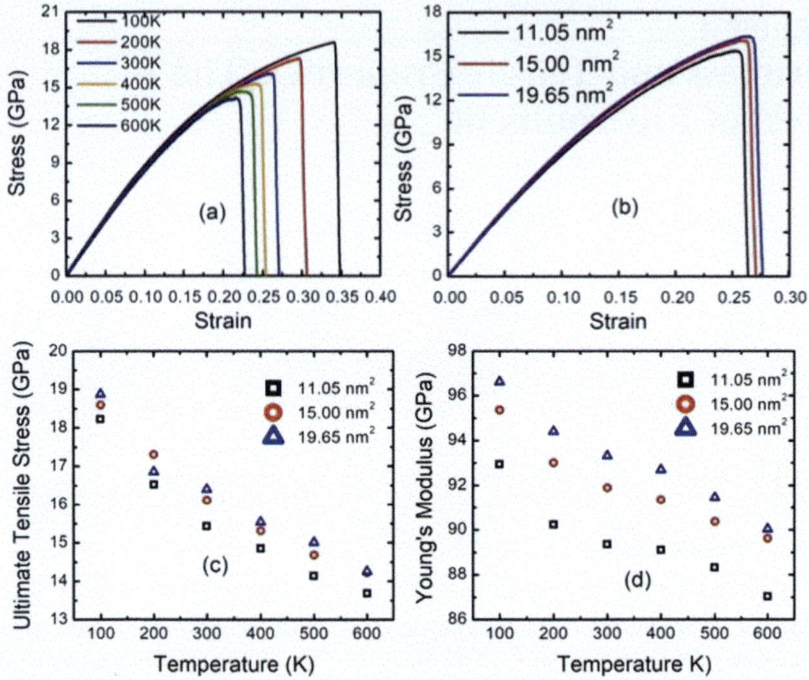

Fig. 4.1 **a** Stress–strain curve of [100]-oriented $Si_{0.5}Ge_{0.5}$ alloy of 15 nm^2 crosssectional area for different temperatures. **b** Stress strain curve of $Si_{0.5}Ge_{0.5}$ for different cross-sectional areas at fixed 300 K. **c** Variations of ultimate tensile stress for different cross-sectional areas of $Si_{0.5}Ge_{0.5}$ alloy with temperature. **d** Variations of Young's modulus for different cross-sectional areas of $Si_{0.5}Ge_{0.5}$ alloy with temperature. Md. H. Rahman, E. H. Chowdhury and Md. M. Islam, J. Nanoparticle Research 22, 311 (2020). With kind permission of Springer Nature

the crack tip. In brittle materials, usually no plastic deformation ahead of the crack tip occurs and the failure sets in instantly due the existing stress concentration at this location. However, failure in $Si_{0.5}Ge_{0.5}$ -which is basically a brittle material– the pattern seems to be different as can be observed in Fig. 4.4. At 100 K (Fig. 4.4a), following cack nucleation at 34.18% strain and its propagation, the effective crosssectional area continuously reduces. At a strain of 34.33%, the cross-section resembles the form of a neck, like a ductile material. At 600 K (Fig. 4.4b), deformation sets in earlier, at a strain of 22.15% strain. Although the alloy is mainly a brittle material, it forms a neck similarly to the observed failure of ductile materials. But unlike to ductile material, where the failure process proceeds relatively slowly, in $Si_{0.5}Ge_{05}$ the neck formation, and failure occur very rapidly.

In summary, the mechanical properties of $Si_{0.5}Ge_{0.5}$ alloy NW was presented in this section. In this predominantly brittle substance neck formation was observed which is a feature of ductile materials.

The presence of oxygen environment oxidizes $Si_{0.5}Ge_{0.5}$ alloy (Zhang et al. 1992). The $Si_{0.5}Ge_{0.5}$ was grown by the molecular beam epitaxy (MBE) on a (100) n-type

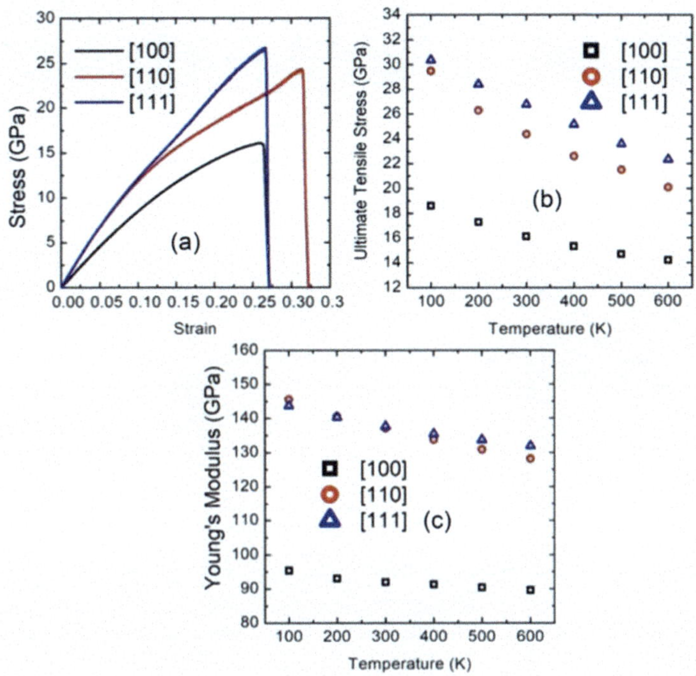

Fig. 4.2 **a** Stress–strain curve of 15 nm^2 $Si_{0.5}Ge_{0.5}$ for different crystal orientation at fixed 300 K. **b** Variations of ultimate tensile stress for different crystal orientation of $Si_{0.5}Ge_{0.5}$ alloy with temperature. **c** Variations of Young's modulus for different crystal orientation of $Si_{0.5}Ge_{0.5}$ alloy with temperature. Md. H. Rahman, E. H. Chowdhury and Md. M. Islam, J. Nanoparticle Research 22, 311 (2020). With kind permission of Springer Nature

silicon substrate. Intentional oxidation was performed using a conventional furnace in dry flowing oxygen at atmospheric pressure at 1000 °C for various times from 1 to 24 h. Rutherford backscattering (RBS) was used to determine the layer thickness (hence growth rates) and the redistribution of germanium. For detailed analysis see the publication of (Zhang et al. 1992).

The variation of the oxide thickness is illustrated in their work, but not discussed here (For details see Zhang et al. 1992). There are two stages during dry oxidation (also in wet oxidation) of the $Si_{0.5}Ge_{0.5}$ alloy at 1000 °C. In the initial period of oxidation, the near-surface layer of oxide consists of SiO_2 and GeO_2. The germanium atoms which are ejected from the oxide, are accumulated at the oxide-alloy interface. The accumulated germanium diffused away from the interface of the oxide layer in the case of dry oxidation and the alloy layer transformed to a germanium-rich layer during wet oxidation. The results demonstrate that the presence of germanium increases the rate during wet oxidation but decreases the rate during dry oxidation.

We explain these phenomena in terms of the mass transport, of either silicon or oxygen atoms, to the oxide front.

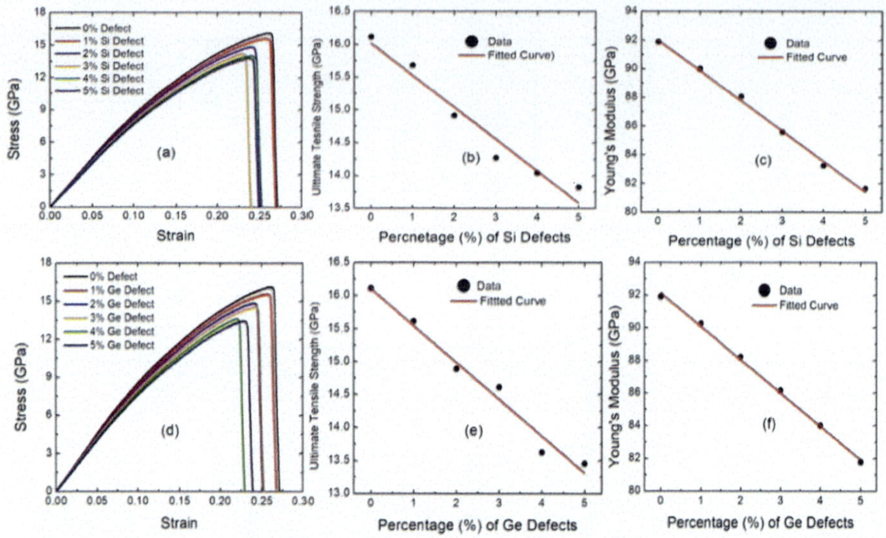

Fig. 4.3 **a** Stress–strain curve of 15 nm^2 [100] oriented $Si_{0.5}Ge_{0.5}$ for different percentage of Si defects at 300 K. **b** Variations of ultimate tensile stress of $Si_{0.5}Ge_{0.5}$ alloy with percentage of Si defects. **c** Variations of Young's modulus of Si0.5Ge0.5 alloy with percentage of Si defects. **d** Stress–strain curve of 15 nm^2 [100] oriented $Si_{0.5}Ge_{0.5}$ for different percentage of Ge defects at 300 K. **e** Variations of ultimate tensile stress of $Si_{0.5}Ge_{0.5}$ alloy with percentage of Ge defects. **f** Variations of Young's modulus of $Si_{0.5}Ge_{0.5}$ alloy with percentage of Ge defects. Md. H. Rahman, E. H. Chowdhury and Md. M. Islam, J. Nanoparticle Research 22, 311 (2020). With kind permission of Springer Nature

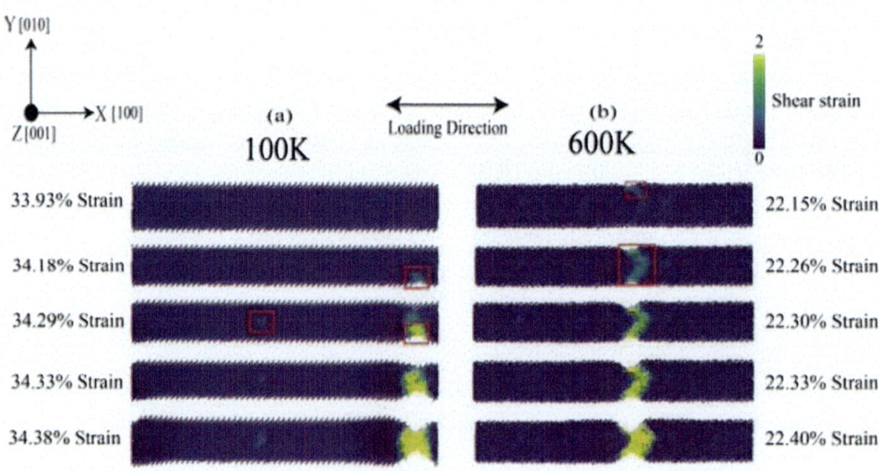

Fig. 4.4 Failure mechanism of Si0.5Ge0.5 at **a** 100K and **b** 600K for various shear strain level. Md. H. Rahman, E. H. Chowdhury and Md. M. Islam, J. Nanoparticle Research 22, 311 (2020). With kind permission of Springer Nature

Compared with bulk silicon the rate of oxidation is reduced for dry oxidation and is similar for wet oxidation. The dominant species are silicon for dry oxidation and oxygen for wet oxidation.

4.2 The Effect of Moisture

4.2.1 In Si

Silicon is a hard, brittle crystalline solid and is a metalloid semiconductor. Silicon is a relatively inactive element at room temperature. It does not combine with oxygen or most other elements. Water, steam, and most acids have very little effect on the element. At higher temperatures, however, silicon becomes much more reactive it combines with oxygen and other elements. A Si based low-k material, SiCOH film was deposited on a p-type silicon substrate by plasma-enhanced chemical vapor deposition (PECVD). The low-k film is porous. A small amount of oxygen was also introduced as an oxidant to remove the organic porogen to form the porous low-k, but specimen without porogen were also prepared. The basic properties are listed in Table 4.1.

Fourier Transform Infrared Spectroscopy (FTIR) was sed for the porous low-k films before and after moisture treatment and is illustrated in Fig. 4.1a. The intensity of H–O-H bonds seems to increase with increasing the immersion time as shown in the inset of Fig. 4.5a. Further, the moisture uptake in the dense and porous low-*k* films, the dependence of the intensity of H–O-H bonds on the moisture immersion time for both low-*k* films is shown in Fig. 4.5b. The importance of immersion time in water is shown in Fig. 4.6 where dense and porous low-k films are illustrated. The hardness and elastic modulus as functions of moisture exposure time are shown for both, dense and porous low-k films. As seen the hardness and modulus slightly increase with increasing the moisture immersion time compared to the as deposited case. However, the mechanical strength in the dense low-k films remained unchanged even after 168 h moisture immersion time. It indicates that the moisture uptake in

Table 4.1 Film properties of the dense and porous low-*k* materials. Yi-Lung Cheng, Ka-Wai Leon, Jun-Fu Huang, Wei-Yuan Chang, Yu-Min Chang and Jihpern Leu. Molecules 2023, 28, 1134. With kind permission of Elsevier

Sample	Precursor	UV curing	Dielectric constant (k)	Breakdown electric filed (MV/cm) at 25°C	Porosity (%)	Pore size (nm)
Dense low-k	DEMS + O_2	No	3.02	9.8	N/D	N/D
Porous low-k	DEMS + O_2 + ATRP	Yes	2.56	6.9	15	1.35

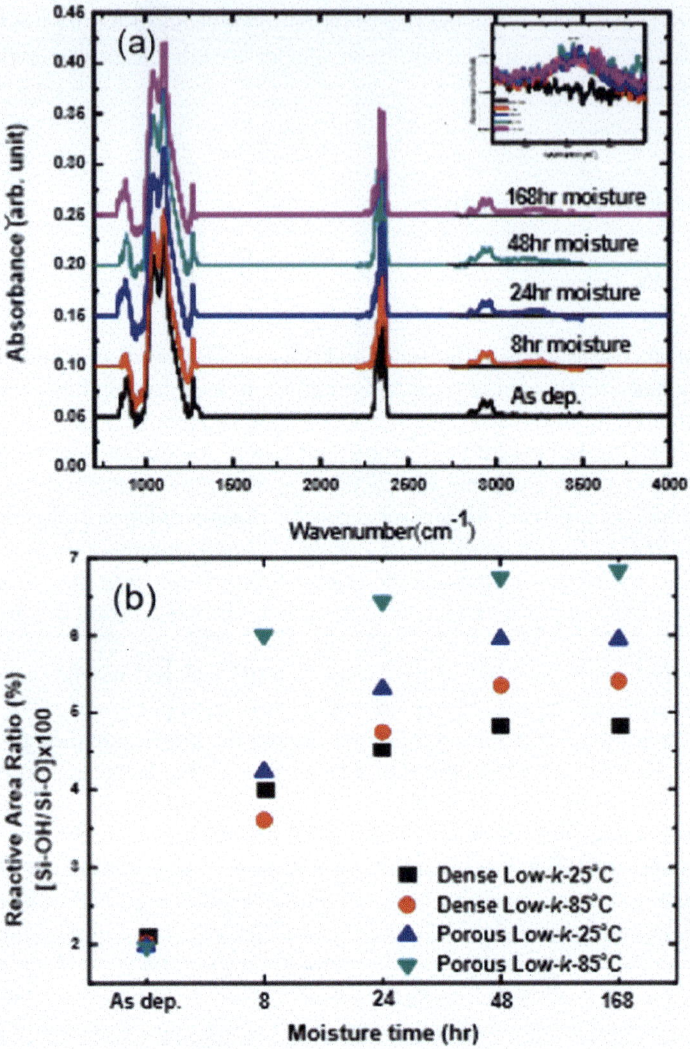

Fig. 4.5 **a** FTIR spectra of low-*k* films after various moisture immersion times. **b** Relative OH-bonding absorbance intensity. Yi-Lung Cheng, Ka-Wai Leon, Jun-Fu Huang, Wei-Yuan Chang. Yu-Min Chang and Jihperng Leu. Microelectronic Engineering, 114, 12 (2014). With kind permission of Elsevier

the low-k porous film improves its mechanical strength. Further, the moisture seems to absorb on the pore wall surface (Fig. 4.7).

In summary, the influence of moisture on the properties of the dense and porous low-k films was presented in this section. It was found that hardness and elastic modulus slightly increase with moisture immersion time and the mechanical strength remained unchanged.

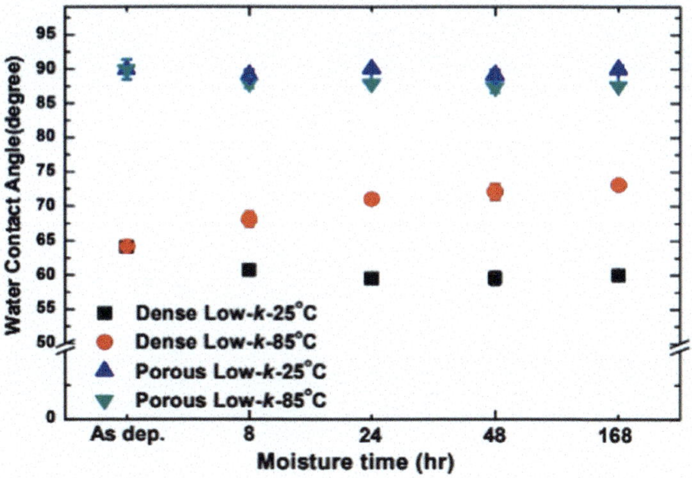

Fig. 4.6 Water contact angle of low-k films after various moisture immersion times. Yi-Lung Cheng, Ka-Wai Leon, Jun-Fu Huang, Wei-Yuan Chang, Yu-Min Chang and Jihperng Leu, Microelectronic Engineering, 114, 12 (2014). With kind permission of Elsevier

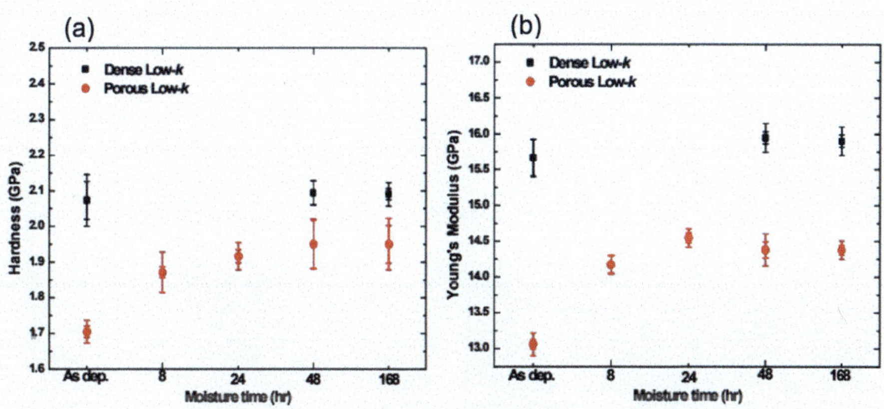

Fig. 4.7 **a** Hardness values **b** elastic modulus as functions of the moisture immersion time for low-k films. Yi-Lung Cheng, Ka-Wai Leon, Jun-Fu Huang, WeiYuan Chang, Yu-Min Chang and Jihperng Leu, Microelectronic Engineering, 114, 12 (2014).With kind permission of Elsevier

References

Y.-L. Cheng, K.-W. Leon, J.-F. Huang, W.-Y. Chang, Y.-M. Chang, J. Leu. Molecules **28**, 1134 (2023)

Md.H. Rahman, E.H. Chowdhury, Md.M. Islam. J. Nanoparticle Res. **22**, 311 (2020)

J.P. Zhang, P.L.F. Hemment, M. Newstead, A.R. Powell, T.E. Whall, E.H.C. Parker. Thin Solid Films **222**, 141(1992)

Chapter 5
Strength—The Effect of Environment

5.1 The Effect of Oxygen

5.1.1 In Si

Silicon is a hard and brittle material widely used in many fields such as in aerospace, military applications, integrated circuits and in many other fields. However, it displays poor preparation methods because their hard and brittle properties at low temperatures. During machining for example damages such as surface micro-crack and subsurface damage might occur. At elevated temperatures say at 700 °C single crystalline Si undergoes plastic deformation and the dislocation generated in the material is accompanied by an abrupt increase in the strength values such as the fracture toughness K_{1c1}. Strength characteristics of single crystal Si is presented in Fig. 5.1 and is compared with Y_2O_3 ceramics. It is presented as a diagram of the two parametric Weibull distribution. The test was carried out at RT and at 700 °C as well (For Y_2O_3 at 800 °C). The measurements of s_c for Si was carried out by four point bending test (For Y_2O_3 three-point bending test). Weibull distribution plots is illustrated in Fig. 5.1 and compared with that of Y_2O_3. In the case of method (a) the fracture toughness, K_{1c} was measured. K_{1c} is given as Eq. (5.1) below as

$$K_{Ic} = Y - 1.5bwP_{2c}L_{0.5}c_k \tag{5.1}$$

where Y is a calculated geometrical constant, c_k the notch depth c being the crack length, w is the same width, L is the spacing of lower supports and l is the distance of upper pressure rolls, b is the width and w the sample thickness. $C = \frac{\Delta v}{\Delta p}$ (v is Poissons ratio), P is the load on the Vickers indenter. In method (b) the value of the fracture toughness was calculated by Eq. (5.2) and is given by

J. Pelleg, *The Influence of Environment on Mechanical Characteristics of Key Semiconductor Components*, https://doi.org/10.1007/978-3-031-34530-2_5

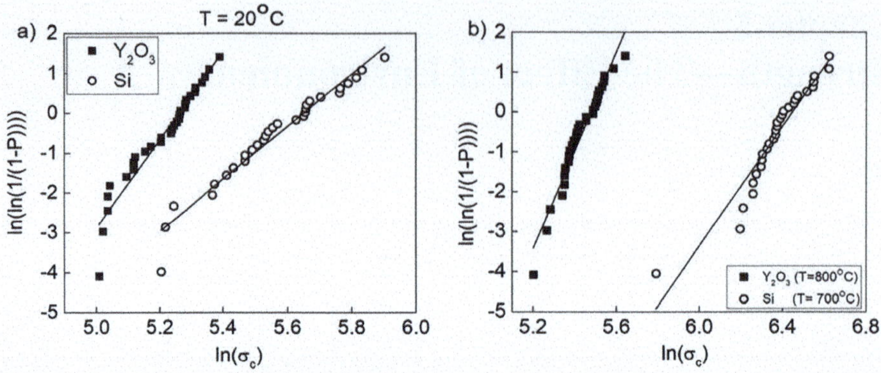

Fig. 5.1 Comparison of the Weibull distribution plots for the strength of the Si single crystal and Y_2O_3 ceramics **a** at room temperature, **b** at elevated temperature. M. Boniecki, P. Kamiński, W. Wesolokovski and K. Krzyżak, Institue of Electronic M1aterials Technology, 44, 4 (2016). Open access

$$K_{Ic} = 0.016\left(\frac{E}{H}\right)^{0.5}\left(\frac{p}{C^{1.5}}\right) \tag{5.2}$$

E is the elastic (Young's) modulus, H is Vickers hardness and P load of the Vickers indenter. As mentioned, s_c for Si was carried out by four-point bending test. For l = 0 (the sample is tested by three-point test, the hardness is given by Eq. (5.3) as

$$H = 1.8544 \times P(2a)^2 \tag{5.3}$$

Λ is the half length of the diagonal of the Vickers impression. And as mentioned P is the load on the Vickers indenter.

For observing the tables of Gaussian and Weibull distributions of the strength of Si single crystals and Y_2O_3 ceramics one should consult Boniecki et al. (2016) of this chapter where the data are assembled in the respective tables. However, the Table on hardness relevant to this section, namely of Si single crystal is presented here as Table 5.1. K_{Ic} and Young's modulus are included in the Table. The oxygen content was ~ 5×10^{15} cm^{-3}.

In summary, hardness data relevant to the subject with oxygen concentration indicated above in Si wafers made from high purity Si single crystal by floating zone (FZ) method, are presented in this section. For the relevant mechanical properties such as bending strength, fracture toughness, Young's modulus, and hardness, one is referred to the original paper Boniecki et al. (2016), where detailed discussion on the experimental methods and the relevant equations are presented in detail.

Table 5.1 Fracture toughness, K_{Ic}. Young's modulus. E and Vickers hardness, H for the single crystal Si at room temperature of 20 °C. M. Boniecki, P. Kamiński, W. Wesolokovski and K. Krzyżak, Institue of Electronic Materials Technology, 44, 4 (2016). Open access

Wafer number	K_{Ic}(MPam$^{1/2}$)	*E* (GPa)	*H* (GPa)
1	0.73 ± 0.07	161 ± 14	9.9 ± 0.8
2	0.79 ± 0.06	172 ± 10	9.6
3	0.78 ± 0.08	178 ± 10	10 ± 1
4	0.70 ± 0.04	180 ± 9	9.8 ± 0.4
Mean	0.75 ± 0.04	173 ± 9	9.9 ± 0.2

5.1.2 *In SiC*

Here instead of SiC, the SiC/SiC composite is presented. It is known to have high specific strength, high specific stiffness, high temperature resistance, long-term oxidation resistance, and erosion resistance, all of which are ideal properties for application as high-temperature-resistant material in many fields such as thermal protection system of aerospace vehicles and hot-end components of aircraft engines. The performance of this composite oxidized in air is important in engineering applications. Pyrolitic carbon (PyC) fiber matrix interface was examined by SEM. The results indicate that PyC interface was severely oxidized and turns into voids above 1000 °C. Oxidation at 1400 °C, the brittle oxide layers filled cracks in the matrix caused drop of tensile strength, which affected the life time of the composite. The results indicate that the tensile strength of the composite decreases with oxidation temperature during 600–1200 °C. Figure 5.2a illustrates the variation of tensile strength and (b) the tensile modulus of SiC/SiC with temperature which was oxidized in air. In Fig. 5.3, similar data are given but oxidized at 1200 °C for different times.

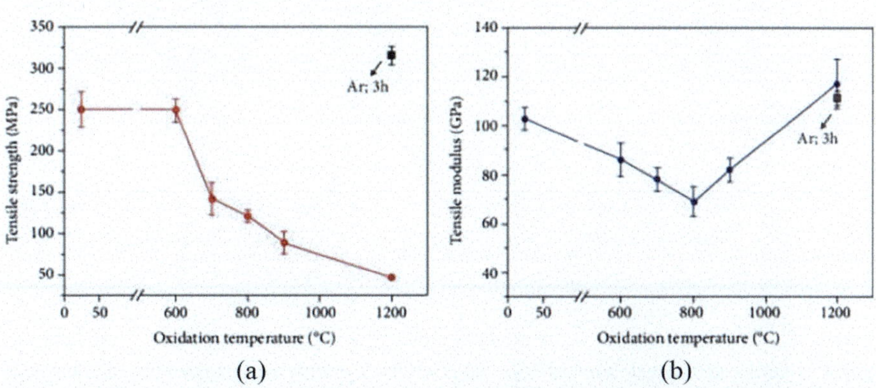

Fig. 5.2 Variation of **a** tensile strength and **b** tensile modulus of SiC/SiC oxidized in air at different temperatures. E. Jin, D. Ma, Z. Yuan, W. Sun, H. Wang, J. Zhang, X. Go1ng, L. Huang, B. Han, X. Sun, Z. Feng, and J. LI, Wiley-Hindawi, 2020, 8,840,963 (2020). Open access

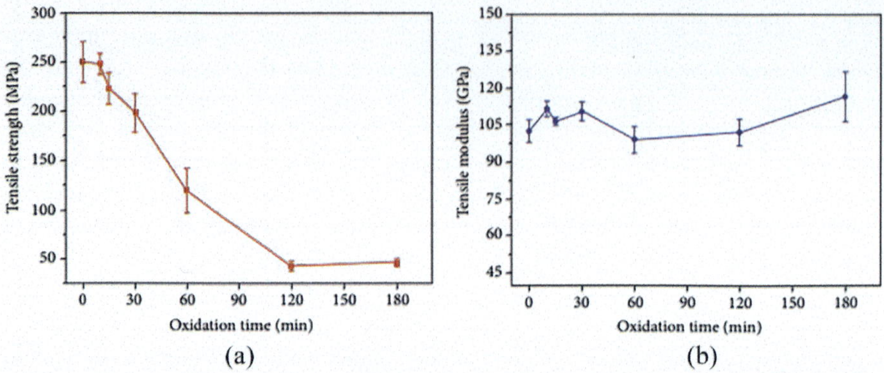

Fig. 5.3 Variation of **a** tensile strength and **b** tensile modulus of SiC/SiC oxidized in air at 1200 °C for different times, E. Jin, D. Ma, Z. Yuan, W. Sun, H. Wang, J. Zhang, X. Go1ng, L. Huang, B. Han, X. Sun, Z. Feng, and J. LI, Wiley-Hindawi, 2020, 8,840,963 (2020). Open access

The relationship between tensile strength loss rate and time oxidized at 1200 °C is shown in Fig. 5.4a and the loss rate and oxidation temperature (3 h oxidation) is shown in Fig. 5.4b. The figures note the calculated loss rate according to Eq. (5.4) and experimental results. Equation (5.4), namely Eq. (5.3) in Fig. 5.4.

$$\begin{aligned} r &= \{0 \;\; (if\ q \le 0) \\ &\{aT\log(t\} + b\;\; (if\ q > 0\} \end{aligned} \tag{5.4}$$

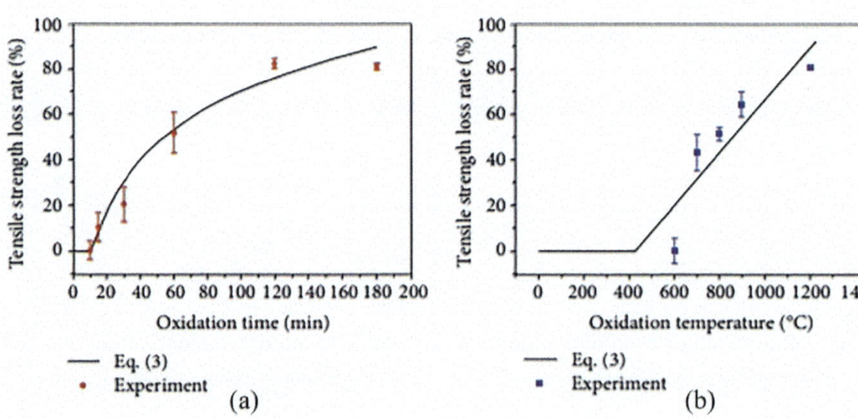

Fig. 5.4 **a** Variation of tensile strength loss rate at 1200 °C for different oxidized time and **b** tensile strength loss rate oxidized at different temperature for 3 h. E. Jin, D. M1a, Z. Yuan, W. Sun, H. Wang, J. Zhang, X. Gong, L. Huang, B. Han, X. Sun, Z. Feng, and J. LI, Wiley-Hindawi, 2020, 8,840,963 (2020). Open access

One gets $a = 0.051$ and $b = -80.41$ by data fitting.

Experimental and calculated loss rate data obtained based on oxidation damage parameter, shown in Fig. 5.4, are in good agreement.

The microstructures at the fracture of SiC/SiC composites are characterized by SEM showing the failure mode and microstructure damages of the fracture can be seen in Fig. 5.5. They illustrate the fracture morphology of SiC/SiC oxidized at 600 °C in air for 3 h. Additional microstructures at other temperature can be seen in Jin et al. (2020)

In summary, the tensile properties of SiC/SiC composite oxidized at high temperatures are presented in this section. Both oxidation temperature and time have significant effects on the tensile properties. The tensile properties begin to decrease with temperature when the oxidation temperature exceeds 600 °C. The tensile modulus decreases with temperature from RT to 800 °C, then increases above 800 °C and crystallization of the matrix occurs. Tensile strength loss rate sets in as observed as a function of temperature and time. Prediction of the results agrees with experimental results. Surface densification treatment can improve tensile strength by reducing oxidation damages of SiC/SiC. For details of additional experimental data obtained in this work, Jin et al. (2020) should be consulted. In the reference many relevant figures and equations ae detailed.

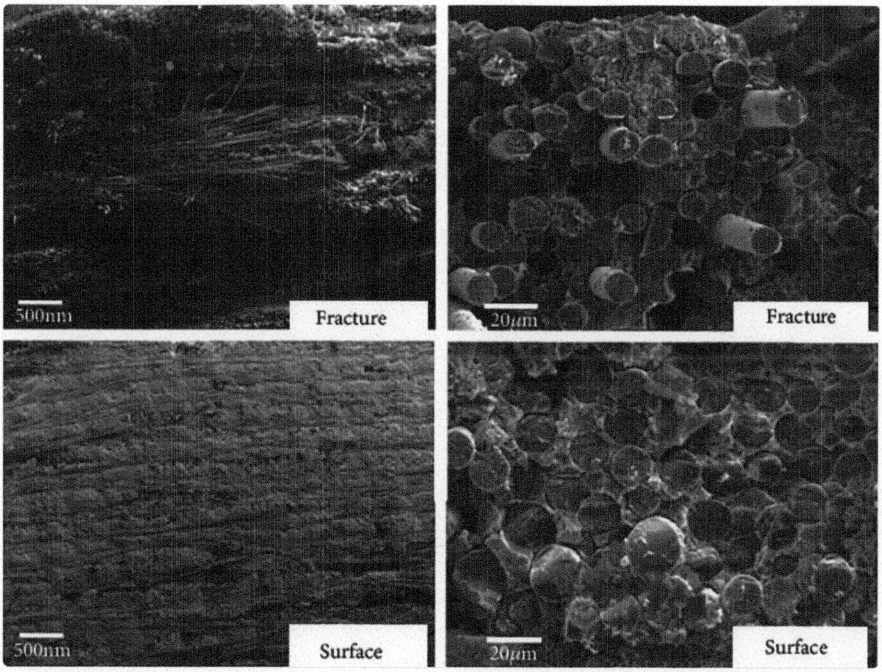

Fig. 5.5 Tensile fracture morphology of Si/SiC composited oxidized in air at 600 °C for 3 h. E. Jin, D. Ma, Z. Yuan, W. Sun, H. Wang, J. Zhang, X. Gong, L. Huang, B. Han, X. Sun, Z. Feng, and J. LI, Wiley-Hindawi, 2020, 8,840,963 (2020). Open access

5.2 The Effect of Moisture

5.2.1 *In SiC*

Tribological tests are carried out on a tribometer (TRB3) in a dry sliding mode at a relative humidity of 50% ± 5% and 90% ± 5% in the ball-on-disc configuration with a load of 15 N. The friction junction consisted of various ball on disk configurations for wear and friction measurements. Of interest in this section is the use of SiC for the tribological test. The friction junction consisted of 100Cr6 steel and SiC. The tribological test results showed reduced linear wear and friction coefficient at a relative humidity of 90% ± 5% compared to the 50% ± 5% humidity junction. The work of a tribological system consists of providing the necessary amount of energy to overcome the resistance resulting from the processes that accompany friction. As a result of friction, the energy is transformed into other forms of energy: thermal, electrical, and mechanical.

Friction implies material losses and energy consumption. The conditions of concern are contact pressure, sliding speed, material type, and environmental factors, such as temperature and humidity. The ball-on-disc tribometer in a dry sliding mode shows in Fig. 5.6 a photograph of the tester and a diagram of the friction junction, while Table 5.2 indicates the process and environmental parameters used. The disc is 100Cr6 steel and Ball is the SiC. During the test the sound was recorded. A sensitron SHT31 sensor was used to measure humidity. To control humidity, an ultrasonic humidifier and a dehumidifier were used. The surface damage after the tests, the geometric structure of the wear tracks was analyzed, and the maximum depth and wear area were determined. For their graphical illustration consult Desaniuk et al. (2021). The parameters of the geometric structure of the surface after the tests is seen in Table 5.3 where all the tests are compared including the junction of interest

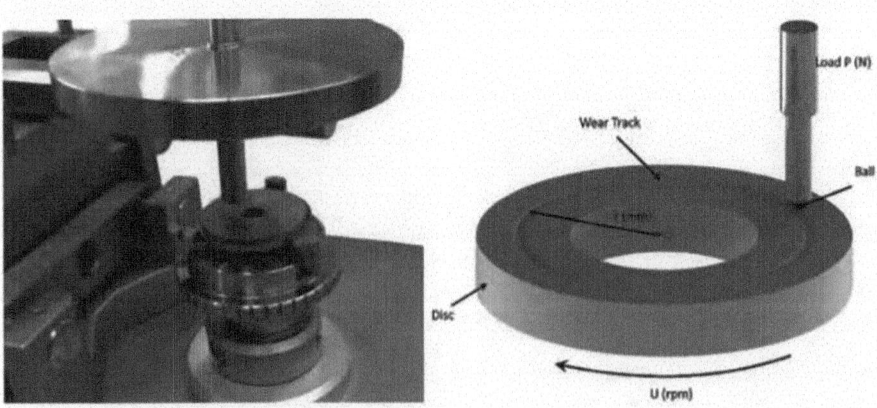

Fig. 5.6 Anton Paar TRB3 tribometer **a**, ball-on-disc diagram, T. Desaniuk, D. Sobon and W. Jurczak, Tribologia 297(3):9–18 (2021). Free access

Table 5.2 Test parameters, T. Desaniuk, D. Sobon and W. Jurczak, Tribologia 297(3):9–18 (2021). Free access

Parameter	Unit	Value
Load	N	15
Sliding rate	m/s	0.07
Number of cycles	–	1000
Humidity	%	50 ± 5; 90 ± 5
Temperature	°C	25 ± 1
Radius	mm	12

of this section, namely 100Cr6 steel and SiC. For the results of coefficient of friction and linear wear as a function of sliding distance for the tribological tests see the illustrations in Fig. 5.4 of Desaniuk et al. (2021). In a bar chart. The average coefficient of friction, the average linear wear area, the average sound level and the average wear are illustrated for all materials tested including that of SiC.

In summary, the friction coefficient and the linear wear were the highest for the 100Cr6–SiC pair. Due to the relative humidity, the friction coefficient, and the linear

Table 5.3 Parameters of the geometric structure of the surface after the tests. T. Desaniuk, D. Sobon and W. Jurczak, Tribologia 297(3):9-18 (2021). Free access

	Sq [μm]	Ssk	Sku	Sp [μm]	Sv [μm]	Sz [μm]	Sa [μm]
100Cr6 steel-100Cr6 stel [Load: 15 N; Humidity: 50% ± 5%]	11.04	– 1.442	3.587	12.74	30.00	42.74	8.674
100Cr6 steel-A1,0$_3$ [Load: 15 N; Humidity: 50% ± 5%]	32.84	0.4357	1.746	65.12	40.54	105.7	28.92
100Cr6 steel-SiC [Load: 15 N; Humidity: 50% ± 5%]	0.7367	– 0.7614	6.818	1.741	4.438	6.179	0.4437
100Cr6 steel-100Cr6 steel [Load: 15 N; Humidity: 90% ± 5%]	2.437	– 0.4676	4.002	8.802	7.670	16.47	1.890
100Cr6 steel-Al,0$_3$ [Load: 15 N; Humidity: 90% ± 5%]	11.94	– 0.465	1.495	14.48	19.82	34.31	11.16
100Cr6 steel-SiC [Load: 15 N; Humidity: 90% ± 5%]	0.6345	0.4933	8.961	2.654	5.044	7.697	0.4235

wear decrease. The sound level is closely related to the performance of the system used, namely, the tribological system.

5.2.2 In GaAs

According to A. J. W. Moore and W. J. Mcg. Tegart Roesch et al. (1994), namely, although hardness has little effect on friction, in general, friction is less with harder materials. It has been shown that the friction decreases linearly with the hardness. For a given load the amount of plastic deformation at the points of intimate contact is greater for a soft metal than for a harder one. So, the subject is indirectly related to some sections of this chapter where hardness and other mechanical properties are discussed in terms of environmental effects. As mentioned friction and wear is a subject of hardness and mechanical properties (Fig. 5.7).

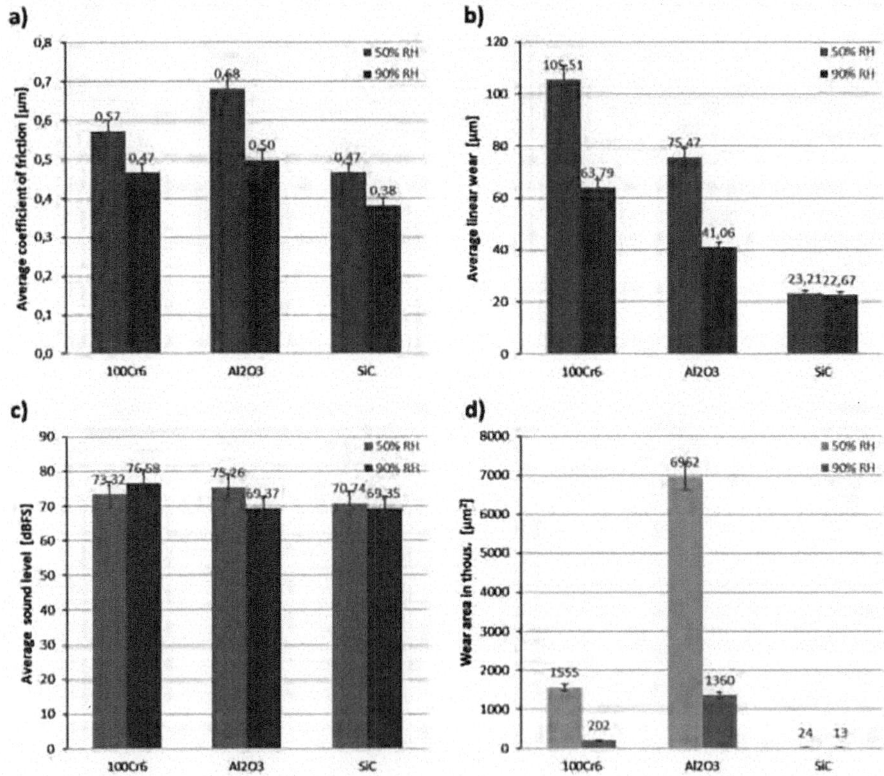

Fig. 5.7 Bar charts: **a** average coefficient of friction, **b** average linear wear, **c** average sound level, **d** wear area. T. Desaniuk, D. Sobon and W. Jurczak, Tribologia 297(3):9-18 (2021). Free access

GaAs devices are generally used in humid environments in hermetic packages. However, in order to reduce the cost of high frequency GaAs devices the use of nonhermetic packages became of interest in the last few years. In order to survive inside non-hermetic packages, GaAs devices must exhibit resistance to humidity. Moisture degradation is a primary concern for operation in non-hermetic packages such as plastic encapsulation. Environmental tests were performed to qualify the reliability of integrated circuits (IC) in such packages (plastic) in order to investigate moisture resistance. The tests performed by a standard testing vehicle and to conduct the test a SPST switch with a built-in driver was applied. The encapsulation is an epoxy mold compound with a glass transition temperature of 160 °C. The humidity tests exceeded the minimum specified 1000 h. All testing was completed without experiencing a failure. Twenty-one 132-pin devices were run in the tests for 1000 cycles without any failure developing. The samples were measured at interim points of 100, 200, and 500 cycles. After completing the test, most of the samples were analyzed for residual gas. No anomalous defects or changes to the die surfaces were observed. Table 5.4 lists the standard test results. The water vapor content found inside each type of sample is shown in Table 5.5. The bias settings are shown in Table 5.6. The changes in FET array disposed to 85% relative humidity at 85 °C is illustrated in Fig. 5.8.

In summary, the results of these tests indicate the resistance of GaAs IC to reliability degradation caused by humidity. Evidence that GaAs devices are usable for

Table 5.4 Standard Test Results, Custom Pin Plastic Package. W. Roesch, R. A. Winters, A. L. Rubalcava and B. Ingle. TriQuint Semiconductor, Inc. Proceedings of 1994 IEEE GaAs IC Symposium

#	Name	JEDEC STD 22 method	Sample size	Failures	Detail
B1	Physical dimentions	B100	2	0	Verify data sheet
B2	Marking permanency	B107	4	0	View @ 5–7 inches, 3X magnification
B3	Solderability	B102	25	0	260 °C, 5 s
B4	Autoclave	A102	100	0	2 atm, 121 °C, 100% RH, 96 h
CI	Lifetest	A108-A	77	0	150 °C, 1500 h, unbiased
C2	Humidity	A101	77	0	85/85,1500 h, unbiased
D1	Lead integrity	A105-A	25	0	8 oz tension and bending stress
D2	Resistance to soldering heat	B106	22	0	260 °C, 10 s
D3	Thermal shock	A106	77	0	– 40 °C to 125 °C, 100 cycles–Liquid
D4	Temperature cycling	A104	77	0	– 40 °C to 125 °C, 100 cycles–Air

Table 5.5 Residual Gas Analysis Results. W. Roesch, R. A. Winters, A. L. Rubalcava and B. Ingle. TriQuint Semiconductor, Inc. Proceedings of 1994 IEEE GaAs IC Symposium

Sample	Type	Moisture (ppm)
#20	Normal	18,100
#23	Normal	18,700
#25	Normal	19,200
#3	Extra epoxy	33,600
#4	Extra epoxy	64,400
#9	Extra epoxy & Water	> 200,000
#10	Extra epoxy and water	> 200,000
#16	Extra epoxy and water	> 200,000

Table 5.6 FET array for Bias Conditions. W. Roesch, R. A. Winters, A. L. Rubalcava and B. Ingle. TriQuint Semiconductor, Inc. Proceedings of 1994 IEEE GaAs IC Symposium. FET stands for field effect transistors

State	Gate (V)	Drain (V)	Source (V)	# of FETs
No bias	0	0	0	12
Idss	0	3	0	9
Reverse	− 2	0	0	16
Pinchoff	− 2	3	0	12

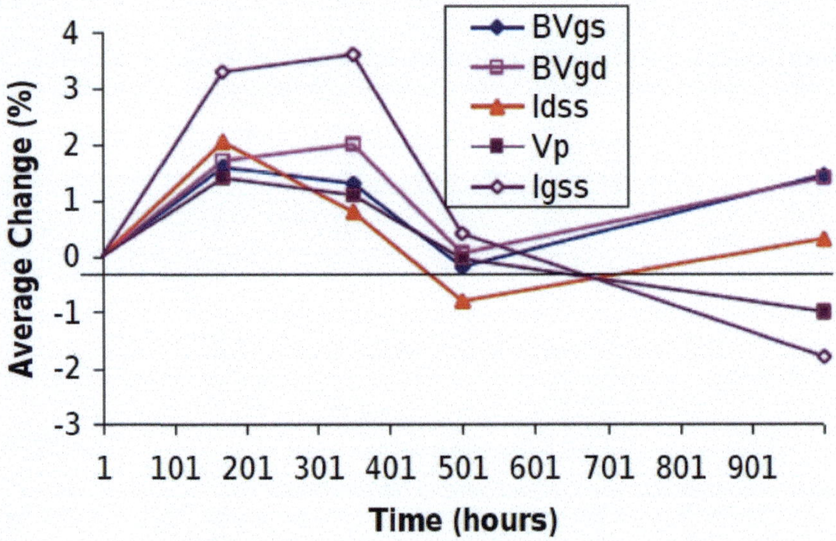

Fig. 5.8 Changes for FET Arrays Directly Exposed to 85°C/85% Relative Humidity. W. Roesch, R. A. Winters, A. L. Rubalcava and B. Ingle. TriQuint Semiconductor, Inc. Proceedings of 1994 IEEE GaAs IC Symposium. FET stands for field effect transistors

the non-hermetic packages is provided and also superior reliability performance compared to silicon devices under humidity conditions is indicated. For detailed experimental and performance analysis, one is directed to Roesch (1994).

The samples were measured at interim points of 100, 200, and 500 cycles. After completing the test, most of the samples were analyzed for residual gas. No anomalous defects or changes to the die surfaces were observed. Table 5.4 lists the standard test results. The water vapor content found inside each type of sample is shown in Table 5.5. The bias settings are shown in Table 5.6. The changes in FET array disposed to 85% relative humidity at 85 °C is illustrated in Fig. 5.8.

References

M. Boniecki, P. Kamiński, W. Wesolokovski, K. Krzyżak. Institue of Electronic Materials Technology, **44**, 4 (2016)

T. Desaniuk, D. Sobon, W. Jurczak. Tribology A **3**, 9 (2021)

E. Jin, D. Ma, Z. Yuan, W. Sun, H. Wang, J. Zhang, X. Gong, L. Huang, B. Han, X. Sun, Z. Feng, J. LI. Wiley-Hindawi, 8840963 (2020)

W. Roesch, R.A. Winters, A.L. Rubalcava, B. Ingle. TriQuint Semiconductor, Inc., in *Proceedings of 1994 IEEE GaAs IC Symposium*

Chapter 6
Brittle Behavior-The Effect of Environment

6.1 The Effect of Oxygen

6.1.1 In SiC

In this section on SiC, SiC/SiC composite is considered. Its high temperature exposure to oxidizing atmosphere introduces substantial changes in mechanical behavior, namely the degradation of some properties as exemplified by flexure tests. The composite contains graphite coated Nicalon reinforced SiC. Recall that Nicalon is a silicon carbide continuous tow that possesses advanced properties when compared to traditional materials, including increased strength and heat and corrosion resistance. Today, products containing Nicalon are mainly used in the aerospace industry as composite reinforcement. The Nicalon reinforced composite is a previously oxidized in air at 900 °C of a SiC/Nicalon composite. Degradation of fracture resistance set in after a very short time of exposure (less than 1 h) also involving fiber pullout. Crack velocities are a function of time and applied stress. Slow crack growth (SCG) controls the long-term life of such aceramic-matrix composite (referred to as subcritical) at elevated temperatures when preexisting flaws are present. Crack bridging plays a crucial role in the mechanical behavior. Environmental effects (for example oxygen) are an important aspect to be considered.

Various bars were exposed to various times at 950 °C and typical loaddisplacement curves for several exposure times are shown in Fig. 6.1. Figure 6.2 show SEM micrographs with decreased amounts of fiber pullout with increasing oxidation time.

In summary, SCG controls the long-term life of Nicalon/0.4 mm graphite/SiC after exposure to air at 950 for various times as illustrated in Fig. 6.2. Crack velocities increase at 1100 °C in the presence of oxygen. A model is developed (He and Anderson 1993) to simulate crack bridged by continuous fiber, (for details see (Jones et al. 1993)). Crack bridging by long fibers plays an important role in the mechanical properties of such materials. Crack bridging reinforcement is highly stressed and may exhibit and a variety of time dependent deformation behavior, such as found

J. Pelleg, *The Influence of Environment on Mechanical Characteristics of Key Semiconductor Components*, https://doi.org/10.1007/978-3-031-34530-2_6

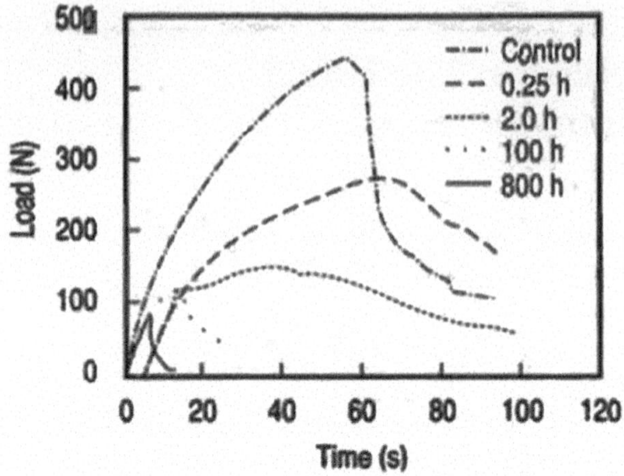

Fig. 6.1 Room temperature flexure curves for Nicalon/0.4 mm graphite/SiC composites after oxidation in air at 950 °C for times shown. R. H. Jones, C. H. Henager, Jr and P. F. Tortorelli, JOM, 45, 26 (1993)

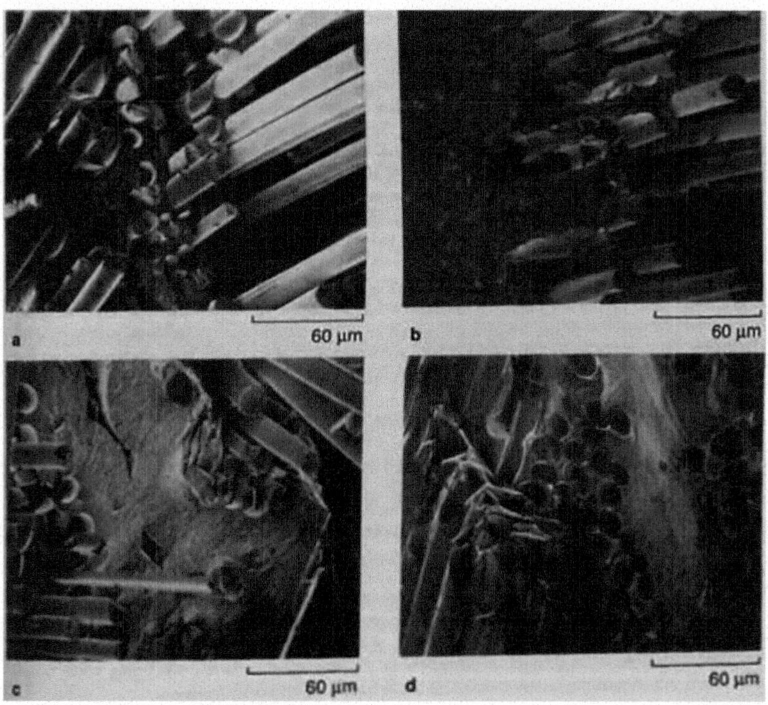

Fig. 6.2 Scanning electron micrographs of fracture surfaces of Nicalon/0.4 mm graphite/SiC after exposure to air at 950 °C **a** 0.25 h, **b** 2 h, **c** 24 h, and **d** 800 h. R. H. Jones, C. H. Henager, Jr and P. F. Tortorelli, JOM, 45, 26 (1993)

for example in creep, in particular when exposed to environmental effects which can induce degradation by changing the nature of the reinforcement-matrix interface.

6.1.2 In GaAs

Oxygen was introduced to GaAs by implantation and investigated by deep level transient spectroscopy (DLTS) and current voltage (I-V) characterization. Contrary to the earlier section on SiC/SiC composite where their brittleness was emphasized, here the electrical properties are investigated. At low temperatures GaAs is brittle, so it was not necessary to emphasize this feature, but rather concentrate on its electrical properties since in integrated circuits and light emitting diodes are of major application.

The post-implantation annealing was studies by RTA (rapid thermal annealing) treatment at 550 °C. The GaAs substrates were two kinds of n and n^* types. At low dose implantation, no trap levels were introduced to the n-GaAs, which means the reduction of carrier concentration. Apparently, this means free carrier trapping. However, at higher dose implantation two new holes were detected in n-GaAs. RTA. treatment showed that the compensation effect (of carier reduction) is cancelled at the 550 °C (Yu et al. 2014). This is believed to occur mainly by damage-related defects. In the case of n^+-GaAs, a higher dose is necessary for equal carrier compensation compared with lower doped n-GaAs. Damage related defects were induced in n^+-GaAs where addition traps were found for 10^{11} and 10^{12} ions/cm^2.

The n and n^+ Si doped GaAs substrates were grown by the Bridgman technique with $N_d = 1 \times 10^{17}$cm^3 and $N_d > 1 \times 10^{18}$/cm^3 (meaning 10^{11} and 10^{12} ions/cm^2 as indicated above). Detailed experimental data can be found in Yu et al. (2014). Table 6.1 summarizes the I-V tests at RT. Typical diode parameters for samples implanted at different doses were calculated from I-V measurements and are summarized in Table 6.2. A nearly ohmic contact was achieved in the highly doped semiconductor with enhanced properties. A rectifying contact could be obtained in a relative highly doped n-GaAs after oxygen implantation. The 10^{10}/cm^2 dose implanted wafer 550 °C RTA, Al-2, showed poor Schottky properties. A nearly ohmic contact was achieved in the highly doped semiconductor case after oxygen implantation. It implies that oxygen implantation may lower surface doping concentration. This change is eliminated by RTA annealing, indicating that implantation damage caused this surface carrier concentration lowering (termed carrier compensation).

Better I-V characteristics were observed after RTA annealing of all higher implanted samples. The improvement is related to the dose itself. The I-V data of the n^+-GaAs is shown in Table 6.3. Due to the very high substrate doping ($N_d > 1 \times 10^{18}$/cm^3), a good Schottky contact is difficult to achieve. However, after implantation the Schottky contact is improved.

DLTS spectra are shown in Fig. 6.3 from the starting GaAs sample A0 and as implanted sample A1-1. As can be seen four electron traps and one hole trap labeled as E1, E2, E3, E4 and H were detected in the starting wafer. The DLTS data are

Table 6.1 Sample identification (GaAs). L. He and W. A. Anderson, Solid-State 36, 173 (1993). With kind permission of Elsevier

Sample	Implantation dose (l/cm^2)/energy (keV)	RTA condition (°C/s)
Si-doped $N_d = 1.07 \times 10^{17}/CM^3$		
A0	None	Starting wafer
Al-I	$10^{10}/50$	As-implanted
Al-2	$10^{10}/50$	550/20
A2-1	$10^{11}/100$	As-implanted
A2-2	$10^{11}/100$	550/20
A3-I	$10^{12}/100$	As-implanted
A3-2	$10^{12}/100$	550/20
Si-doped $N_d > 10^{18}/cm^3$		
B2-1	$10^{11}/100$	As-implanted
B2-2	$10^{11}/l00$	550/20
B3-1	$10^{12}/100$	As-implanted
A3-2	$I0^{12}/100$	550/20

Table 6.2 I-V data for oxygen implanted n-GaAs. L. He and W. A. Anderson, Solid-State Electronics, 36, 173 (1993). With kind permission of Elsevier

Sample	Leakage current J_0 (A/cm^2)	Barrier height $Ø_B$ (eV)	Ideality factor n
A0	4.52×10^{-3}	0.478	–
A1-1	1.63×10^{-6}	0.698	1.53
A1-2	3.22×10^{-3}	0 441	–
A2-1	5.54×10^{-9}	0.846	1.15
A2-2	8.22×10^{-3}	0.476	–
A3-1	8.92×10^{-8}	0.773	2.42
A3-2	3.27×10^{-5}	0.620	3.23

Table 6.3 I-V data for oxygen implanted n$^+$-GaAs. L. He and W. A. Anderson, Solid-State Electronics, 36, 173 (1993). With kind permission of Elsevier. $N_d > 1 \times 10^{18}/cm^3$

Sample	Leakage current J_0(A/cm^2)	Barrier height $Ø_B$(eV)	Ideality factor n
B2-1	3.57×10^{-4}	0.558	2.22
B2-2	1.71×10^{-5}	0.637	1.70
B3-1	3.64×10^{-7}	0.737	1.44
B3-2	2.34×10^{-7}	0.748	1.38

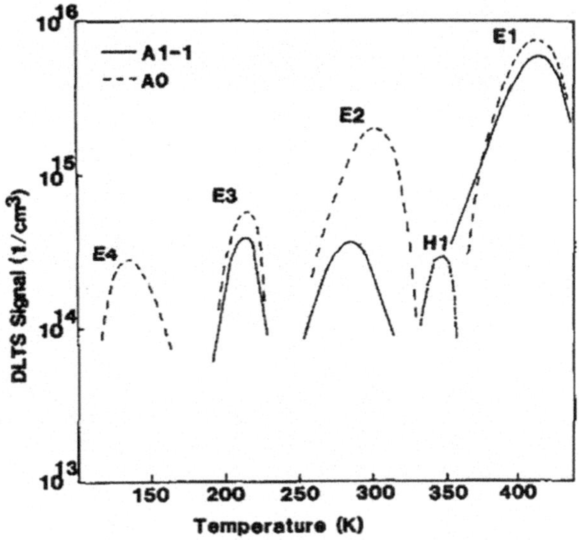

Fig. 6.3 DLTS spectra of n-GaAs samples from starting wafer A0 and implanted wafer A1-1. L. He and W. A. Anderson, Solid-State Electronics, 36, 173 (1993). With kind permission of Elsevier

summarized in Table 6.4. The electron traps were found to have activation energies of 0.82, 0.54, 0.43 and 0.12 eV, respectively, while the hole trap H1 was found with E, = E, + 0.76 eV. Arrhenius plots of the thermal emission rates of the deep levels in n-GaAs are seen in Fig. 6.4. The RTA post-implantation annealed sample A1-2 showed similar DLTS spectrum as the implanted sample A1-1, however it gave higher trap concentration for E1 and E3. Figure 6.3 shows DLTS spectra from n-GaAs samples at different implantation energies and doses. All three samples shown are as implanted and show great differences. Figures 6.4 and 6.5 show DLTS spectra from n-GaAs with oxygen, implantation at energy 100 keV and doses 10^{11} and 10^{12} from both as-implanted as post-implantation RTA treated samples.

DLTS measurement was also performed for oxygen implanted n^+-GaAs samples. Figures 6.6 and 6.7 illustrate them. They were implanted at two different doses of $10^{11}/cm^2$ and $10^{12}/cm^2$ with 100 keV implantation energy. The differences between the two cases are very obvious. Two new electron traps, E5 and E6 appeared in the 10^{11} dose implanted sample at low temperature with activation energies below the conduction band by 0.31 eV and 0.17 eV. After RTA treatment traps E1 and E3 and a hole trap H1 were observed, but the shape of the spectrum was distorted.

In summary, compared with n-GaAs, higher dose implantation is required in n^+-GaA for compensation purposes. Additional traps were detected in oxygen implanted n^+- GaAs, which was damage induced defect related. For higher dose ($10^{12}/cm^2$) implanted n-GaAs two more-hole traps were detected which could form induced acceptor levels. Compared with n-GaAs, higher dose implantation is requiring in n^+- GaAs for compensation purposes. The figures are self-explanatory, but for their full and detailed description one is referred to the original publication in He and Anderson (1993).

Table 6.4 DLTS parameters of this work. L. He and W. A. Anderson, Solid-State Electronics, 36, 173 (1993). With kind permission of Elsevier. [15] of the Table is listed as [10]

Trap labeled	Activation energy E_a (eV)	Capture section $\sigma(cm^2)$	Trap density N_1 (cm^{-3})
E1	$E_c - 0.80 - 0.82$	$(5.2 - 8.0) \times 10^{-14}$	$> 6 \times 10^{15}$
E2	$E_c - 0.48 \sim 0.54$	$(1.8 - 2.9) \times 10^{-14}$	$> 2 \times 10^{15}$
E3	$E_c - 0.36 \sim 0.43$	$(2.0 - 3.2) \times 10^{-14}$	$> 2 \times 10^{14}$
E4	$E_c - 0.12$	3.9×10^{-18}	2×10^{14}
E5	$E_c - 0.31$	2.3×10^{-18}	2×10^{16}
E6	$E_c - 0.17$	1.5×10^{-16}	1×10^{15}
H1	$E_v + 0.76$	1.8×10^{-13}	4×10^{14}
H2	A broad band		
H3	$E_v + 0.76$	2.7×10^{18}	1×10^{15}
Work of others			
EL2(a)	$E_c - 0.31$	$(0.8 - 1.7) \times 10^{-13}$	[15]
EL3(b)	$E_c - 0.31$	$(0.8 - 1.7) \times 10^{-13}$	
EL5(b)	$E_c - 0.31$	$(0.5 - 2.0) \times 10^{-13}$	
EB10(c)	$E_c - 0.31$	Imprecise	
EL16(d)	$E_c - 0.31$	4.0×10^{18}	
EL11(e)	$E_c - 0.31$	3.0×10^{16}	
G(f)	$E_c - 0.31$	2.4×10^{12}	[16]

(a)Arsenic antisite vacancy; (b) Point defect/impurity complexes; (c) Unknown; (d) Acceptor-like defects; (e) An impurity; (f) Native defects

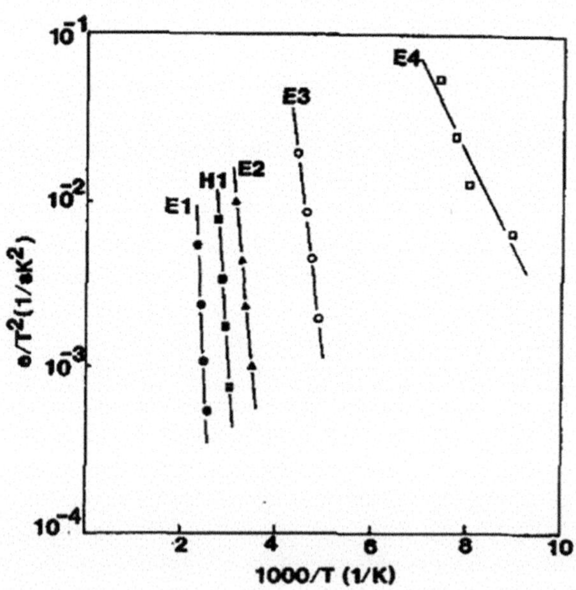

Fig. 6.4 Arrhenius plots of emission rates from the traps found in Fig. 6.1. L. He and W. A. Anderson, Solid-State Electronics, 36, 173 (1993). With kind permission of Elsevier

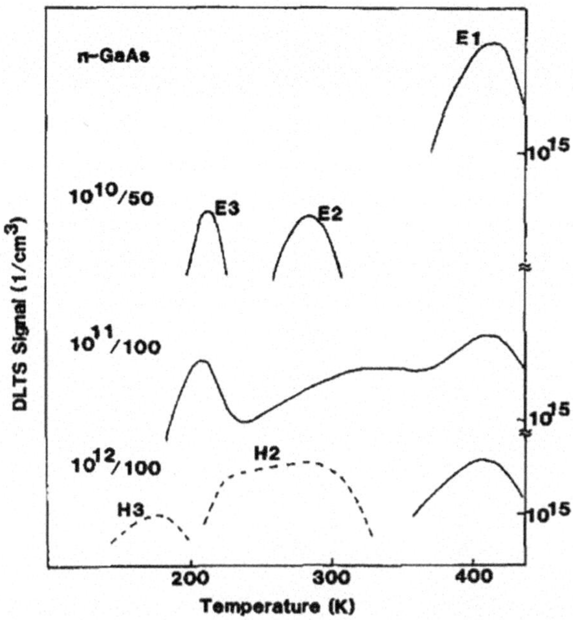

Fig. 6.5 DLTS spectra of n-GaAs samples with different implantation conditions: **a** Al-1, $10^{10}/cm^2$, 50 keV; **b** A2-1, $10^{11}/cm^2$, 100 keV; **c** A3-1, $10^2/cm^2$, 100 keV. Solid line represents electron traps and dashed line is for hole traps. L. He and W. A. Anderson, Solid-State Electronics, 36, 173 (1993). With kind permission of Elsevier

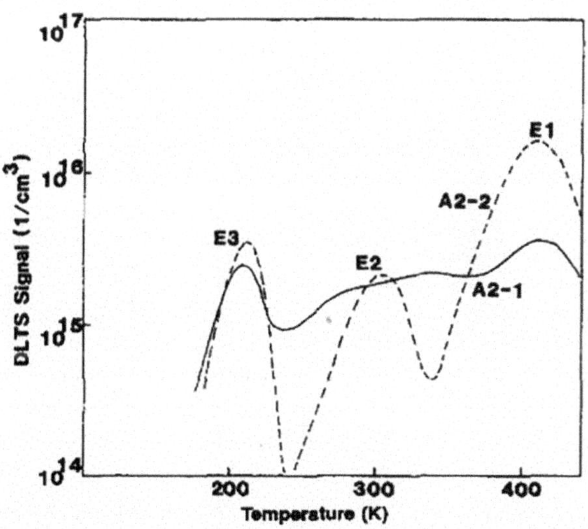

Fig. 6.6 DLTS spectra of $10^{11}/cm^2$, 100 keV implanted n-GaAs as-implanted sample A2-1 and RTA treated sample A2-2. L. He and W. A. Anderson, Solid-State Electronics, 36, 173 (1993). With kind permission of Elsevier

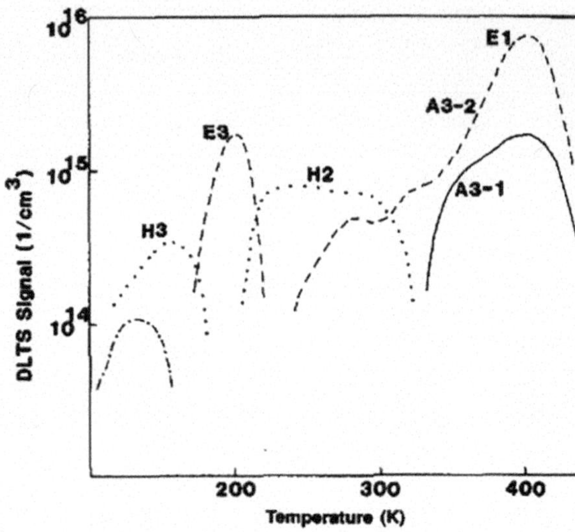

Fig. 6.7 DLTS spectra of $10^2/cm^2$, 100 keV implanted n-GaAs as-implanted sample A3-1 and RTA treated sample A3-2. L. He and W. A. Anderson, Solid-State Electronics, 36, 173 (1993). With kind permission of Elsevier

6.2 The Effect of Moisture

6.2.1 In SiC

The application of SiC presented in this section is related to the influence of humidity on the SiC performance. It considers the wear and friction characteristics of SiC, namely its tribological performance. SiC is an advanced structural ceramic, exhibiting high hardness and elastic modulus, low thermal expansion, and good thermal stability. At ambient temperature, SiC is exceptionally stable and do not react with surrounding atmosphere.

When two ceramics such as SiC (a pair) slide against each other, mechanical processes such as indentation, abrasion and attrition produce a large number of tiny debris with an increased surface area. SiC ceramics have excellent structural properties, such as high hardness and melting point, high strength, good chemical resistance, good electrical conductivity, and excellent wear resistance. Because of these qualities SiC ceramics are useful for various tribological applications[8]. However, under unlubricated sliding contacts they exhibit poor friction characteristics, which can greatly deteriorate the durability and reliability of a tribosystem. The tribological performance can be improved by controlling the microstructure by reducing grain size of SiC ceramics, acquiring elongated SiC grains and reducing the amount of grain boundary intergranular phase etc.. For additional improvement consult Ref. 8. The objective of this section is to present SiC design with high wear resistance and low friction.

In Table 6.5 a summary of different grain sizes and shapes are presented collected from the literature while Fig. 6.8 presents the grains in SiC with the additions indicated which are sintering additives. AlN-Sc_2O_3 additives induces large, elongated grains and equiaxed grain formation and show low wear in SiC. Note, that the large, elongated grains in SiC ceramics mean more tortuous crack path, compared with equiaxed grains which can result in higher fracture toughness. However, higher wear was observed with amorphous intergranular phase. With these additives amorphous interphase is induced, but no effect on the friction coefficient. The amorphous interface is seen in Figs. 6.9a and in b clean grain boundary is shown.

Less wear resistance of SiC ceramics with the addition 3 wt% Y_2O_3 was caused by the presence of rich weak amorphous phase (namely Y_2O_3) existing along the grain boundaries as a consequence of which microcrack induced fracture formation occurred. Intergranular phases with slip lines can be observed in SiC ceramics with the addition of 3 wt% Y_2O_3 as seen in Fig. 6.10, whereas transgranular fracture can be observed in SiC ceramics with the addition of 0.2 wt% Y_2O_3 because of less secondary phase is present and that resulted in direct bond between SiC grains, leading to improvement in wear resistance. Therefore, it can be concluded that better mechanical properties (hardness and strength) are obtained.

As indicated above, amorphous grain boundary phases are harmful to tribological properties of SiC ceramics at room temperature, but sintering additives of Al, B and C can fully crystallize amorphous SiC in situ at temperatures > 1100 °C leading to improved fracture toughness at 1200–1300 °C. This prior heat treatment may reduce the harmful influence of amorphous grain boundaries on the tribological properties at room temperature. Sintering in N_2 environment leads to improved hardness of the intergranular phase. Thus, the improvement of the tribological performance of SiC is a result of the amorphous phase crystallization and/or the hardening of the intergranular phase. Data on sliding friction and wear is shown in Table 6.6.

The wear resistance of SiC can be also improved by controlling crack propagation. It is believed that wear in monolithic SiC is attributed to the microfracture and grain pull–out at a load of 20 N. In monolithic SiC only crack deflection is observed, but with WC particles addition crack deflection and bridging is seen as illustrate in Fig. 6.11. The sliding friction and wear results of SiC with different characteristics of cracks are summarized in Table 6.7.

The above part on SiC ceramics in unlubricated sliding has been added in order to compare (from the same communication [11]) the effect of humidity as one of the external factors on the tribological performance of SiC. It is reported, that tribochemical reactions can be accelerated when SiC is subjected to sliding under humid conditions. Indeed, humidity is an important factor in the tribology of SiC. It was found that wear debris formed during the initial stage of sliding when a SiC slid at any relative humidity of 20%, 50% and 90%. During increasing sliding distance debris are continuously pulled out during the sliding at low humidity level of 20% causing high friction and severe wear. However, the formed wear debris can be aggregated by water vapor to form a protective layer with the increasing sliding distance when the humidity level is high as 50% and 90%. This can be seen in Fig. 6.12. The protective coating is mainly composed of $Si(OH)_4$. The reaction of the process is given as

$$SiC + O_2 + H_2O \rightarrow SiO_2 + CO + H_2 \text{ and}$$

Table 6.5 Sliding friction and wear results of SiC ceramics with different grain sizes and shapes. W. Zhang, S. Yamashita and H. Kita, Materials and Deign, 190, 108,528 (2020). Open access. (For the listed references see the original communication)

Materials	Sintering additives/the secondary phases	Amount (wt%)	(Primary) SiC size (μm)	Shape of SiC grains	Counterbody	Load (N)	Sliding speed (m/s)	Temperature (°C)	Relative humidity (*)	COF	Specific wear rate (mm3/ N-m)
Si/SiC disk [13]	Free Si	12	45	Equiaxed	Si/SiC pin	20	0.45	15	30–50	0.48	–
Si/SiC disk [13]	Free Si	12	88	Equiaxed	Si/SiC pin	20	0.45	15	30–50	0.47	–
Si/SiC disk [13]	Free Si	12	45	Equiaxed	Si/SiC pin	20	0.45	300	–	***032***	–
Si/SiC disk [13]	Free Si	12	88	Equiaxed	Si/SiC pin	20	0.45	300	–	0.63	–
Si/SiC disk [13]	Free Si	12	45	Equiaxed	Si/SiC pin	20	0.45	600	–	0.47	–
Si/SiC disk [13]	Free Si	12	88	Equiaxed	Si/SiC pin	20	0.45	600	–	0.60	–
SiC disk [15]	WC	0	0.835	Equiaxed	SiC ball	20	0.16	27 ± 5	40 ± 10	0.48	1.3×10^{-5}
SiC disk [15]	WC	30	0.578	Equiaxed	SiC ball	20	0.16	27 ± 5	40 ± 10	0.47	8.0×10^{-6}
SiC bar [18]	AIN + SCjO,	3	–	Thick, largely elongated and fine equiaxed	SiC ball	13	0.21	20 ± 5	50 ± 10	0.23	4.4×10^{-5}

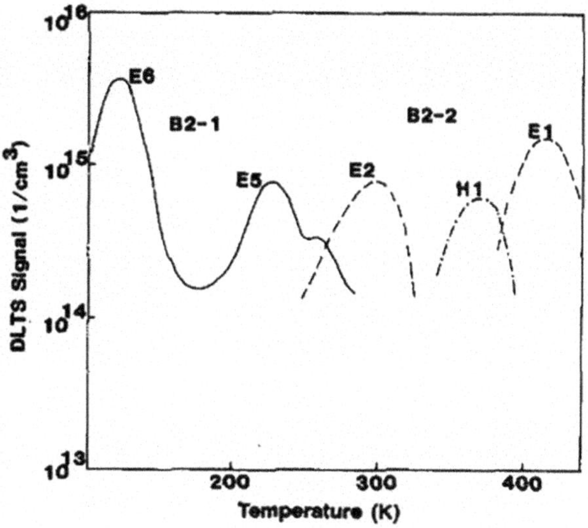

Fig. 6.8 DLTS spectra of $10^{11}/cm^2$, 100 keV implanted n^+- GaAs as-implanted sample B2-1 and RTA treated sample B2-2. L. He and W. A. Anderson, Solid-State Electronics, 36, 173 (1993). With kind permission of Elsevier

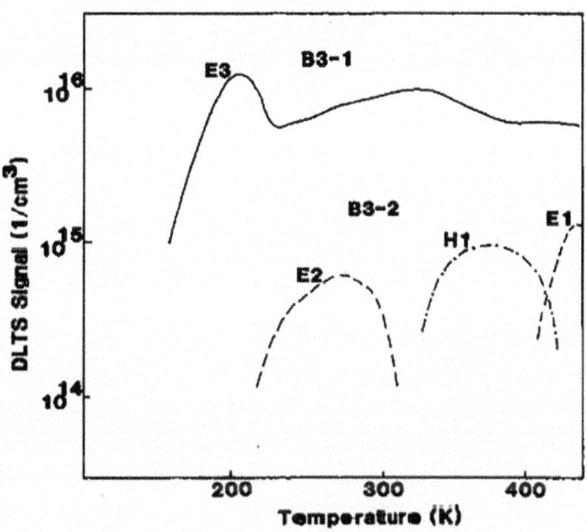

Fig. 6.9 DLTS spectra of $10^{12}/cm^2$, 100 keV implanted n + -GaAs, as-implanted sample B3-1 and RTA treated sample B3-2. L. He and W. A. Anderson, Solid-State Electronics, 36, 173 (1993). With kind permission of Elsevier

$$SiO_2 + 2H_2O \rightarrow Si(OH)_4$$

Table 6.8 summarizes the sliding friction and wear in SiC at various relative humidity, while Table 6.9 shows the effect of temperature at some relative humidity. Water can accelerate crack growth, on the other hand, the earlier mention protective layer formed owing to tribochemical reactions can protect worn surface. As a whole, crack growth and tribochemical reactions compete with each other. For SiC ceramics, tribochemical reactions dominate wear behavior with the increased humidity, which

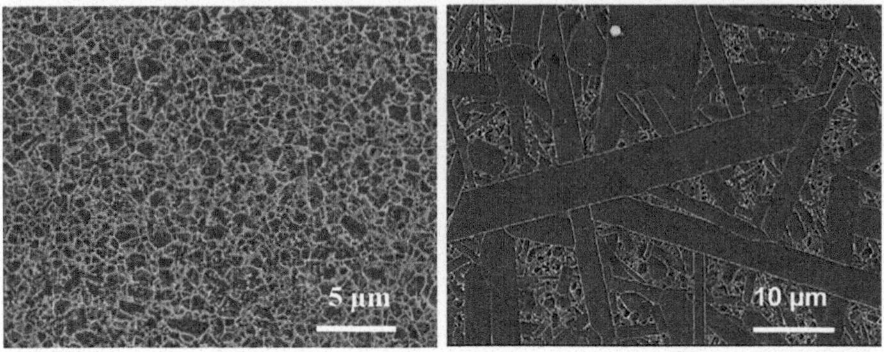

Fig. 6.10 Grains in the SiC ceramics added with **a** WC [15], and **b** AlN and Sc_2O_3 [18]. W. Zhang, S. Yamashita and H. Kita, Materials and Deign, 190, 108,528 (2020). Open access. (For the listed references see the original communication)

can inhibit the wear caused by microfracture, leading to the reduced wear rate with the increasing humidity. However, it is expected that this protection supported by the protective layer against microfracture would loss effectiveness based on imposed conditions of sliding, such as sliding speed, load, and surface roughness, as long as the protective layer is damaged (Fig. 6.13).

For additional information regarding the effect of atmosphere on friction and wear, one should consider inspecting the original communication of Zhang et al. (2020) where a detailed analysis of the experimental procedure and the results are described.

6.2.2 In GaAs

For the effect of humidity regarding the resistance of GaAs IC (integrated circuit) against degradation, a series of investigations of acceptable humidity tests were performed that has shown evidence that GaAs devices are ready for low-cost nonhermetic packages applications and have superior reliability performance compared to silicon devices. This observation is important to manufacturers for using low-cost packaging solutions. Here a complete set of environmental stress tests are used to qualify the reliability of integrated circuits in plastic packages (low-cost) and evaluate the moisture resistance. The qualification tests of low-cost packages provide a basis for reliability investigation. Standard tests do not usually provide insight into types of failure mechanisms, failure distribution or activation energies. Moisture degradation is a primary concern for plastic encapsulation and for the system operation in the nonhermetic packages.

In the standard testing performed according to the Joint Electron Device Engineering Council (JEDEC) (known currently as Solid-State Technology Association) the encapsulation was in an epoxy having a glass transition temperature of 160 °C.

Table 6.6 Sliding friction and wear results of SiC ceramics with different characteristics of grain boundaries. W. Zhang, S. Yamashita and H. Kita, Materials and Deign, 190, 108,528 (2020). Open access

Materials	Sintering additives/the secondary phases	Amount (wt %)	Characteristics of grain boundaries	Counterbody	Load (N)	Sliding speed (m/s)	Temperature (°C)	Relative humidity (*)	COF	Specific wear rate (mm^3/N-m)
SiC bar [81]	AIN + Y_2o_3	3	Amorphous phase	SiC ball	13	0.21	20 ± 5	50 ± 10	0.24	5.3×10^{-5}
SiC bar [18]	AIN + Sc_2o_3	3	Clean	SiC ball	13	0.21	20 ± 5	50 ± 10	0.23	$4\,0.4 \times 10^{-5}$
SiC bar [18]	AIN + Sc_2o_3	10	Clean	SiC ball	13	0.21	20 ± 5	50 ± 10	0.22	5.7×10^{-5}
SiC disk [30]	Y_2O_3	0.2	Less secondary phase	WC ball	20	0.08	27 ± 5	35 ± 10	0.45	2.9×10^{-6}
SiC disk [30]	Y_2O_3	3	Large amount of amorphous phase	WC ball	20	0.08	27 ± 5	35 ± 10	0.44	6.7×10^{-6}

(a) (b)

Fig. 6.11 TEM images of SiC ceramics revealing **a** amorphous grain boundary with the sintering additives of Al-Y_2O_3 and **b** clean grain boundary with the sintering additives of Al-Y_2O_3. W. Zhang, S. Yamashita and H. Kita, Materials and Deign, 190, 108,528 (2020). Open access. (For the listed reference see the original communication)

Of the various tests performed, the humidity test is of interest here. In an unbiased humidity test, the moisture is allowed to saturate the plastic and penetrate completely to the die surface. Whereas in a biased test, the heat that is generated on the circuit tends to drive the moisture away. The GaAs metallization system is primarily gold and corrosion are less likely than when IC interconnections are formed of aluminum. The test conducted was 1500 h and both life and humidity tests exceeded the minimum specified 1000 h. All testing was completed without experiencing a failure. The results indicate that the 5000-ppm moisture limit set for hermetic packages is not applicable to GaAs devices. Instead, in the test, a combination of extra water and extra epoxy (which can outgas water vapor under some conditions) were purposely introduced into the package cavity prior to seal. Fr the detailed experimental consult (Yu et al. 2014). Devices were run in this test for 1000 cycles without any failures developing. The samples were measured at interim points of 100,

200, and 500 cycles using as a product testing sequence. The devices that had extra water and epoxy inside the package could not be measured since they swamped the instrument with moisture, but it was estimated that at least 20% moisture was inside the package. The samples were then examined optically at high power. No anomalous defects or changes to the die surfaces were observed.

For continued reading on this subject and of the work by the same authors in the same article go back to chapter 5 where more details on the subject are provided.

Further data of the effect of humidity in GaAs is presented below. A model based on the reactive tip scratching-induced oxidation, water-solubility of debris and adhesion effect was proposed to interpret the tribochemical removal of GaAs surface. Gallium arsenide (GaAs) serves as an excellent host for optoelectronic devices, because of its excellent physical properties such as direct bandgap and high electron mobility. It is one of the substrates for new generation of quantum structures

Table 6.7 Sliding friction and wear results of SiC ceramics with different characteristics of cracks. W. Zhang, S. Yamashita and H. Kita, Materials and Deign, 190, 108,528 (2020). Open access

Materials	Sintering additives/the secondary phases	Amount (wt %)	Characteristics of cracks	Counterbody	Load (N)	Sliding speed (m/s)	Temperature (°C)	Relative humidity (%)	COF	Specific wear rate (mm^3/N-m)
SiC disk [15]	WC	0	Deflection	SiC ball	20	0.16	27 ± 5	40 ± 10	0.48	1.3×10^{5}
SiC disk [15]	WC	10	Deflection and briding	SiC ball	20	0.16	27 ± 5	40 ± 10	0.45	9.3×10^{6}
SiC disk [15]	WC	30	Deflection and briding	SiC ball	20	0.16	27 ± 5	40 ± 10	0.47	8.0×10^{6}
SiC disk [15]	WC	50	Deflection and briding	SiC ball	20	0.16	27 ± 5	40 ± 10	0.40	4.0×10^{6}

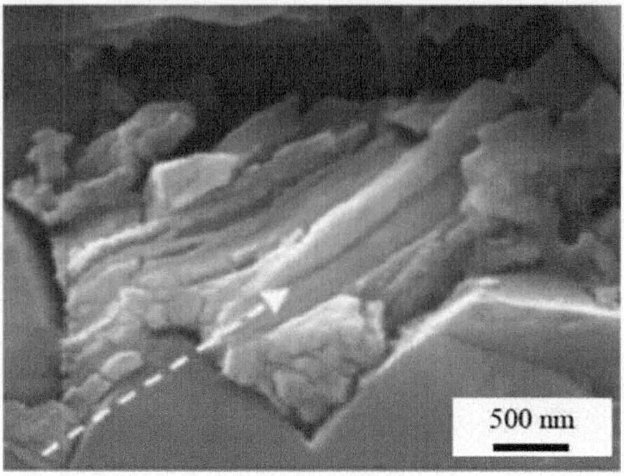

Fig. 6.12 SEM of worn surface of SiC ceramics with the addition of 3 wt% Y2O3. W. Zhang, S. Yamashita and H. Kita, Materials and Deign, 190, 108,528 (2020). Open 1access

Table 6.8 Sliding friction and wear results of SiC ceramics under different relative humidity. W. Zhang, S. Yamashita and H. Kita, Materials and Deign, 190, 108,528 (2020). Open access

Materials	Counterbody	Load (N)	Temperature (°C)	Sliding speed (m/s)	Relative humidity (%)	COF	Specific wear rate (mm^3/N-m)	Specific wear rate of balls (mm^3/N-m)
SiC disk [130]	SiC ball	30	19 ± 2	0.002	20	0.65	6.2×10^{-4}	–
SiC disk [130]	SiC ball	30	19 ± 2	0.002	50	0.3	2.6×10^{-4}	–
SiC disk [130]	SiC ball	30	19 ± 2	0.002	90	0.18	92×10^{-5}	–
p-SiC disk [131]	a-SiC ball	1	Ambient	0.036	30	0.8	0.67×10^{-6}	2.11×10^{-6}
p-SiC disk [131]	u-SiC ball	1	Ambient	0.036	45	0.3	0.26×10^{-6}	0.16×10^{-6}
P-SiC disk [131]	a-SiC ball	1	Ambient	0.036	60	025	0.18×10^{-6}	0.07×10^{-6}
SiC disk [132]	SiC ball	1	Ambient	1	5	0.55	–	–
SiC disk [132]	SiC ball	1	Ambient	1	25	0.6	–	–
SiC disk [132]	SiC ball	1	Ambient	1	50	1.08	–	–
SiC disk I132]	SiC ball	1	Ambient	1	80	12	–	–

Table 6.9 Sliding friction and wear results of SiC ceramics under different temperatures. W. Zhang, S. Yamashita and H. Kita, Materials and Deign, 190, 108,528 (2020). Open access

Materials	Countcrixxly	load (N)	Sliding speed (m/s)	Temperature (X]	Relative humidity {%)	COF	Specific wear rate (tnm!/N ■ m)
SiC coating [49]	SiC ball	2	0.1	25	50–60	0.43–0.48	4.68 × 10 5
SiC coating [49]	SiC ball	2	0.1	600	–	0.65–0.75	Without obvious wear
SiC coating [49]	SiC ball	2	0.1	800	–	05–0.6	151 × 10 3
SiC disk [127]	SiC ball	4	0.1	20–400	–	0.45	-
SiC disk [127]	SiC ball	4	0.1	400–700	–	0.16	-
SiC disk [127]	Al2o3 ball	4	0.1	20–400	–	0.65	-
SiC disk [127]	Al2o3 ball	4	0.1	400–700	–	0.4	-
SiC disk [128]	SiC ball	19	0.078	28 ± 2	55 ± 5	0.47	13 × 10 5
SiC disk [128]	SiC ball	19	0.078	500	–	> 1.0	1.6 × 10" 5
SiC-30 wt%WC [128]	SiC ball	19	0.078	28 ± 2	55 ± 5	0.38	0.3 × 10 s
SiC-30 wt%WC [128]	SiC ball	19	0.078	500	–	> 1.0	1.1 × 10 5
SiC disk [126]	SiC ball	5	0.67	− 196	–	0.4	2a × 10 6
SiC disk [126]	SiC ball	5	0.89	− 196	–	036	0 0.4 × 10 '
SiC disk [126]	SiC ball	5	1.1	− 196	–	033	1.0 × 10 b
SiC disk [126]	SiC ball	5	33	− 196	–	035	1.6 × 10 *
SiC disk [126]	SiC ball	10	33	− 196	–	030	6.0 × 10 b
SiC disk [126]	SiC ball	20	***33***	− 196	–	0.28	83 × 10"*

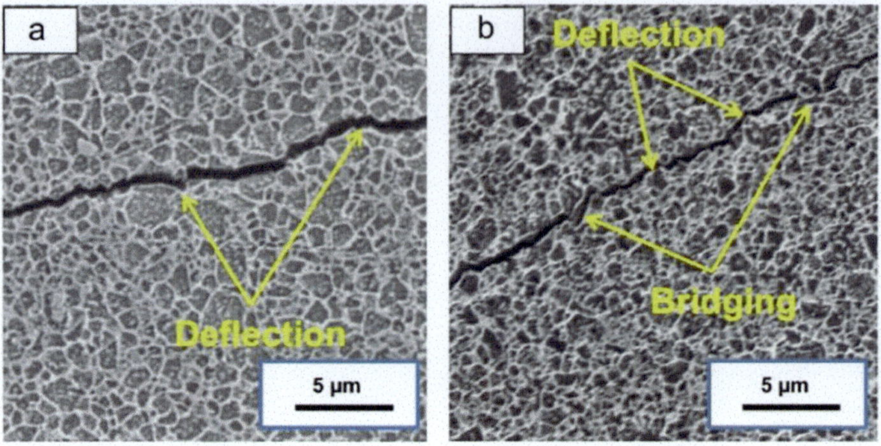

(a) (b)

Fig. 6.13 Crack propagation in: **a** monolithic SiC ceramics and **b** SiC-30 wt.% WC ceramics. W. Zhang, S. Yamashita and H. Kita, Materials and Deign, 190, 108,528 (2020). Open access

for solar cells and laser devices. Especially, patterned GaAs substrate provides an effective way for forming nanoscale structures and upgrading the performance of optoelectronics devices. However, the defects such as dislocations can be introduced during patterning substrates and will degrade the optical properties. For improving the performance of GaAs-based device, it is of great significance to fabricate patterned GaAs substrates without introducing crystal defect (Figs. 6.15 and 6.16).

Recently, tribochemical removal by reactive tip scratching on GaAs surface provides an effective method to produce defect-free substrate. Different from the plastic deformation induced by mechanical cutting with a diamond tip, the material removal can be realized on GaAs surface by a SiO_2 tip with the contact pressure less than the yield pressure of GaAs. The motivation to investigate GaAs is a consequence of the observation of tribochemical reaction of SiO_2/Si pair, that the adsorbed water on silicon surface plays a key role in the tribochemical removal of silicon surface. To investigate the humidity-dependent material removal on GaAs, which may provide new insight into the understanding of the tribochemical process and facilitates the site-controlled fabrication including nano fabrication. The surface root-mean-square (RMS) roughness of the GaAs surface was measured as less than 0.5 nm over a 2 μm $\times$ 2 μm area. A SiO2 tip with a tip radius R of 1.25 μm was used for the nano scratch test. The relative humidity RH ranging from 1 to 90% was controlled by adjusting the ratio of H_2O vapor and dry air in *atomic force microscopy* (AFM) chamber, which is used for in situ investigations. For the scratching, the applied normal load F_n was 2 μN, the sliding velocity, v, was set as 10 μm/s, and the number of scratching cycles N was 100. For comparison, the scratching was also performed in vacuum (~1 Pa) and in deionized water under the same loading conditions. In addition to a SiO_2 tip ($R =$ 600 nm), a diamond tip ($R = 600$ nm) was also used for the test of adhesive force

F_a. All AFM tests were conducted under a temperature of about 25 °C. The wear of the SiO2 tip, was used to detect the tip topography before and after the scratch tests under N = 2000, $F_n = 2\ \mu N$ and relative humidity (RH) = 50%. The cross-section microstructure of the scratched area after material removal under RH = 50% and $F_n = 2\ \mu N$ was observed by a TEM. The AFM images of the scratches on the GaAs surface is illustrated in Fig. 6.14. When the scratch tests were performed in humid air, the relative humidity (1–90%) had a strong effect on the material removal of GaAs surface. The removal of material becomes more and more obvious with the increase in relative humidity. Shallower scratches were produced either in vacuum or in dry air, while much deeper grooves were produced in water. The variation of the groove depth and width is plotted as a function of the humidity in Fig. 6.17.

These results clearly indicate that the existence of water has a great impact on the material removal on GaAs. The material removal in GaAs was performed by SiO_2 tip and diamond tip. The results are shown in Fig. 6.18.

In summary the defect-free tribochemical removal of GaAs was realized under vacuum, dry air and various humidity by scratching with a SiO_2 tip. The removal depth increases with the humidity (1–90%) and reaches its maximum value in water.

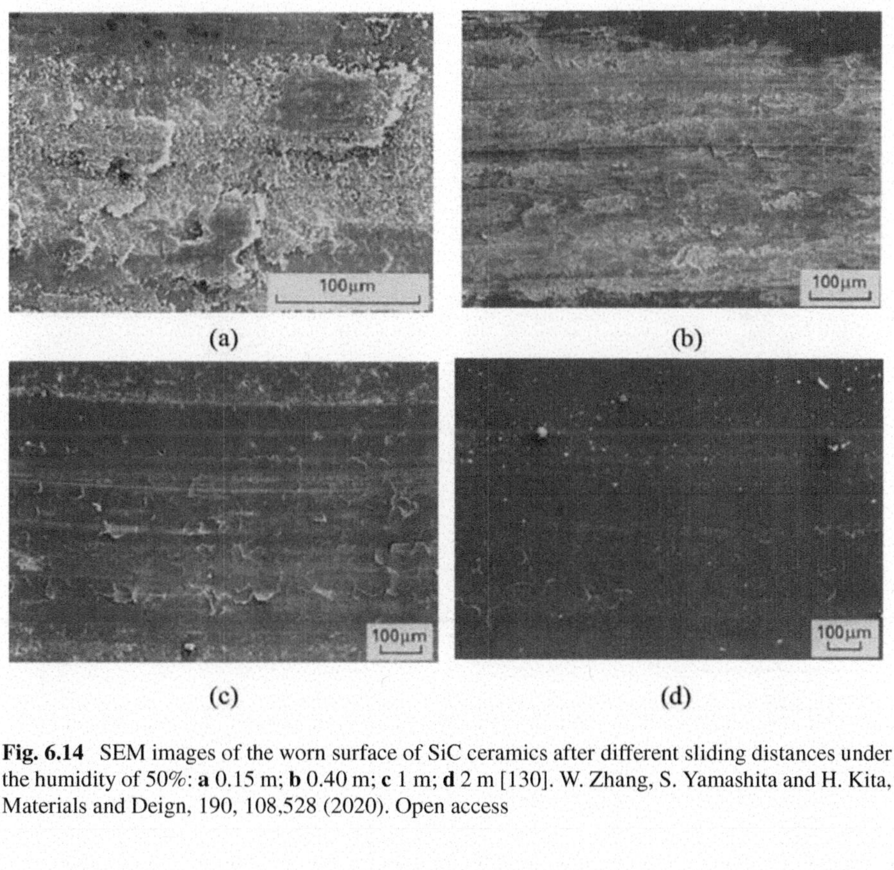

Fig. 6.14 SEM images of the worn surface of SiC ceramics after different sliding distances under the humidity of 50%: **a** 0.15 m; **b** 0.40 m; **c** 1 m; **d** 2 m [130]. W. Zhang, S. Yamashita and H. Kita, Materials and Deign, 190, 108,528 (2020). Open access

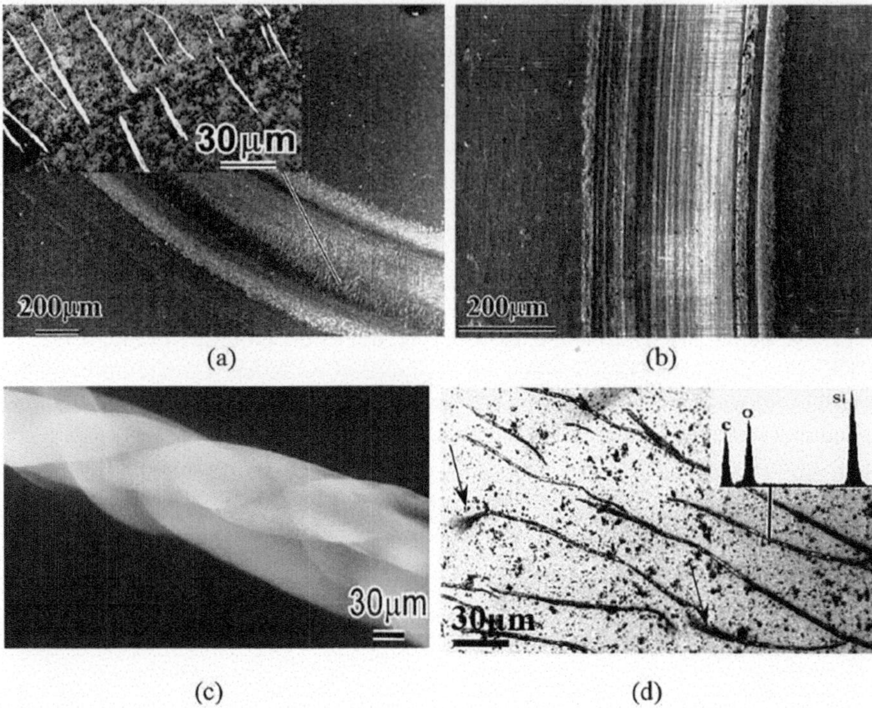

Fig. 6.15 Wear tracks of SiC ceramics after sliding at **a** 30% RH, **b** 60% RH. **c** Twinned silica gel particle and **d** 'brush-like' ends. W. Zhang, S. Yamashita and H. Kita, Materials and Deign, 190, 108,528 (2020). Open access

No obvious damage was observed on the SiO_2 tip before and after the wear test. Perfect crystal matrix was found on the cross section of the scratched groove on GaAs by TEM.

The SiO_2 tip is proved to be intact after scratching, and a controllable fabrication of surface grooves can be realized on GaAs surface under a given scratching condition. The tribochemical removal is a new route to produce patterned GaAs substrate for the fabrication of site-controlled quantum dots and nanowires for high-quality devices.

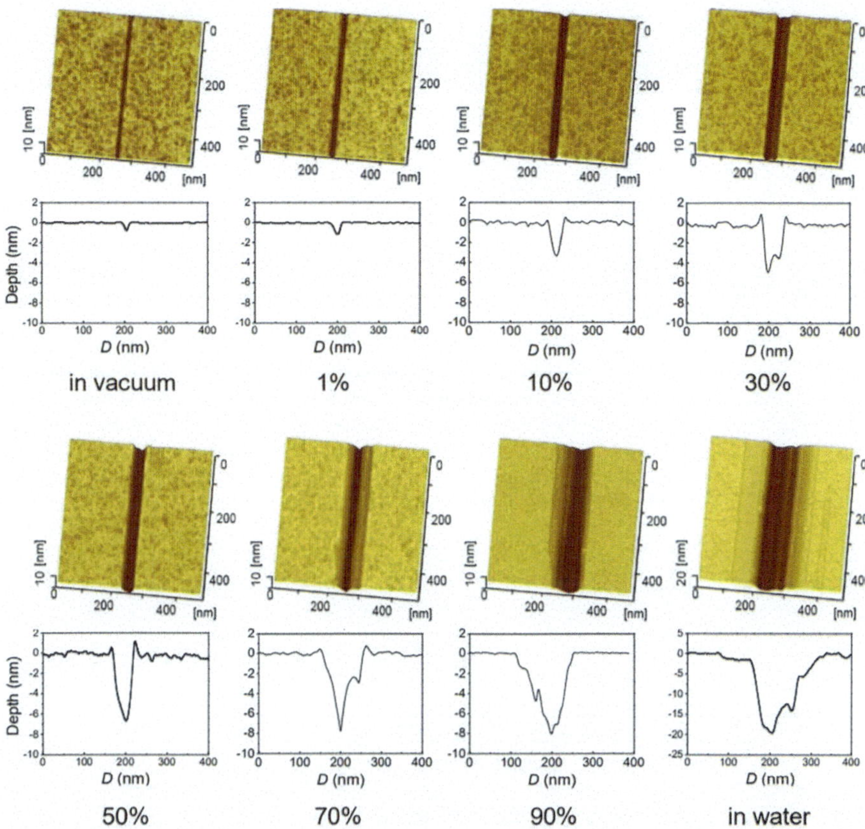

Fig. 6.16 AFM images (top) and the corresponding profiles (bottom) of the scratches on GaAs surface scratched by the SiO_2 tip in vacuum, water, and various relative humidity. B. Yu, J. Gao, C. Jin, C. Xiao, J. Wu, H. Liu, S. Jiang, L. Chen and L. Qian, Template for APEX (Jan. 2014). Also Applied Physics Express, 9, 066,703 (2016). Open access

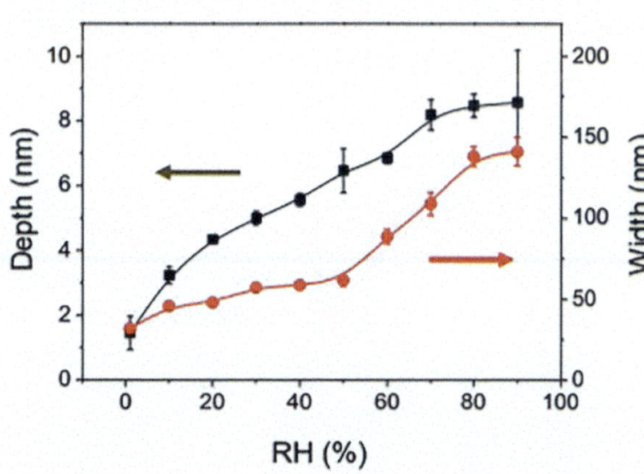

Fig. 6.17 Variation of the groove depth and width as a function of relative humidity. B. Yu, J. Gao, C. Jin, C. Xiao, J. Wu, H. Liu, S. Jiang, L. Chen and L. Qian, Template for APEX (Jan. 2014) and more specifically in Applied Physics Express, 9, 066,703 (2016). Open access

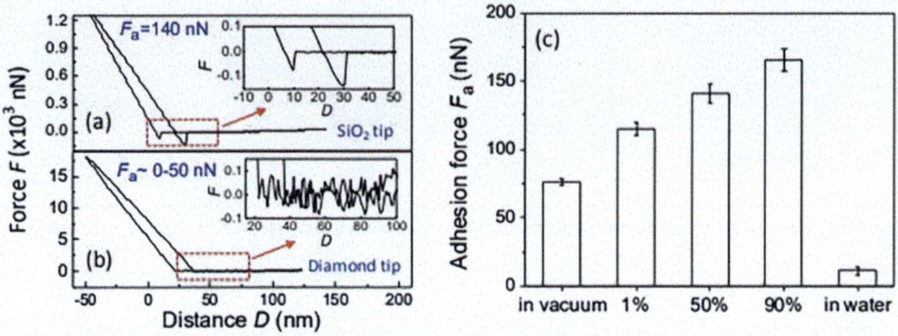

Fig. 6.18 The adhesive force F_a measured on GaAs surface by a SiO_2 tip (R = 600 nm) and by a diamond tip (R = 600 nm), respectively. Under RH = 50%, **a** F_a of about 120 nN is measured by the SiO_2 tip, while **b** F_a measured by the diamond tip is too low and submerged in the noise peaks. **c** F_a measured by the SiO_2 tip under vacuum, water, and various humidity. B. Yu, J. Gao, C. Jin, C. Xiao, J. Wu, H. Liu, S. Jiang, L. Chenand L. Qian, Template for APEX (Jan. 2014) and more specifically in Applied Physics Express, 9, 066,703 (2016). Open access

References

M. Boniecki, P. Kamiński, W. Wesolokovski, K. Krzyżak. Institue of Electronic Materials Technology, **44**, 4 (2016)

T. Desaniuk, D. Sobon, W. Jurczak. Tribologia **3**, 9 (2021)

L. He, W.A. Anderson. Solid-State Electron. **36**, 173 (1993)

E. Jin, D. Ma, Z. Yuan, W. Sun, H. Wang, J. Zhang, X. Gong, L. Huang, B. Han, X. Sun, Z. Feng, J. LI, CaG

R.H. Jones, C.H. Henager Jr, P.F. Tortorelli. JOM **45**, 26 (1993)

A.J.W. Moore, W.J. Mcg. Tegart, in *Proeedings of Royal Soceity A,* **212**, 452 (1952)

W. Roesch, R.A. Winters, A.L. Rubalcava, B. Ingle. TriQuint Semiconductor, Inc. in *Proceedings of 1994 IEEE GaAs IC Symposium* (1994)

B. Yu, J. Gao, C. Jin, C. Xiao, J. Wu, H. Liu, S. Jiang, L. Chen, L. Qian. Template for APEX (Jan 2014)

W. Zhang, S. Yamashita, H. Kita, Mater. Deign **190**, 108528 (2020)

Chapter 7
Creep–Time Dependent Deformation. Effect of Environment

7.1 The Effect of Oxygen

7.1.1 In SiC

SiC based ceramic matrix composites (CMCs) are utilized for their refractory properties in the aerospace industry. The composition and structure of these materials are crucial to maintaining the strength, toughness, oxidation, and creep resistances that are desired of silicon carbide. Six different morphologies present in the silicon carbide matrix were observed: smooth, nodular, rough nodular, bumpy, nucleated, and plate-like. It was found that high matrix tensile strength minicomposite batches contained solely the smooth morphology, while low-matrix tensile strength minicomposite batches contained a variety of other morphologies (Tobin et al. 2021). Smooth SiC is the only lift-out that does not have significant oxygen character in the matrix. Smooth SiC also solely exhibits the (111) planes of cubic SiC, while the other morphologies display all four intense peaks from cubic SiC. Thus, the randomly oriented crystal structure and the presence of oxygen are contributors to the weaker strength of those morphologies compared to smooth SiC.

The materials considered here are minicomposites composed of Hi-NICALON SiC fibers, a boron nitride (BN) chemical vapor infiltrated (CVI) interphase, and an atmospheric CVI-applied SiC matrix. Methylrichlorosilane (MTS, CH_3SiCl_3, 98%) was used. Ultrahigh-purity nitrogen and ultra-high purity hydrogen were also required. A visual representation of this process is displayed in Fig. 7.1. For all illustrations, experimental procedure and for detailed analysis consult (Tobin et al. 2021) of this chapter. However, the main point has been stated above, namely, the presence of oxygen is detrimental since a reduction in strength has been observed.

In Table 7.1 of the work of Wilshire and Burt (Wilshire and Burt 2015) four SiC reinforced composites are indicated. The interest for this section lies in the standard SiC_f/SiC, where Nicalon is the reinforcement. The interface is carbon < 0.5 mm. The creep and creep fracture properties of the 0/90° fiber-reinforced CMCs

J. Pelleg, *The Influence of Environment on Mechanical Characteristics of Key Semiconductor Components*, https://doi.org/10.1007/978-3-031-34530-2_7

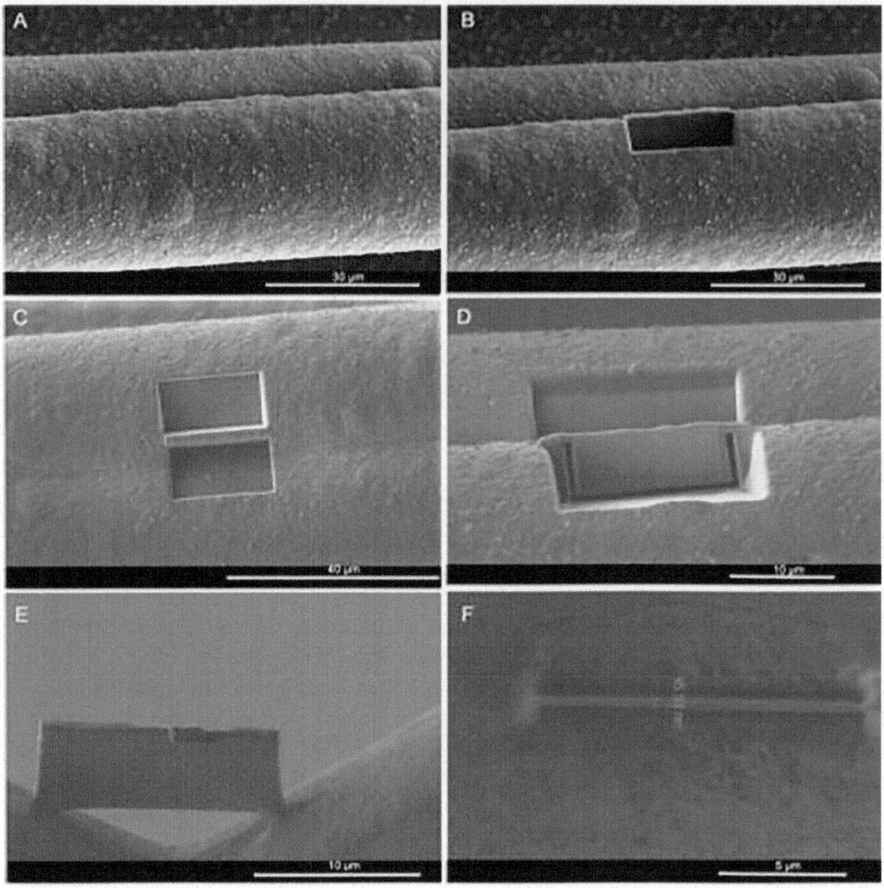

Fig. 7.1 Focused ion beam (FIB) lift-out process of the SiC matrix from SiC_f/SiC minicomposites. **a** Platinum cap is deposited on desired area. **b** Gallium ions mill out on side of lamella. **c** Gallium ions mill out both sides of lamella. **d** U-cut of a rectangle in the lamella. **e** Rectangle cut from lamella is loaded onto the TEM grid. **f** Finished lamella is thinned to TEM thickness (~100 nm). Z. Tobin, K. Petroski, P. Kerns, A. Almansour, J. Grady and S. L. Suib, ACS Omega, 6, 863 (2021). Open access

can be compared using data obtained in tensile creep tests carried out in air at 1573 K. Continuously decaying creep curves were recorded, i.e., after the initial strain on loading at the creep temperature, the creep rate decreased gradually with time towards a minimum value, with no welldefined secondary or tertiary creep stages apparent before fracture occurred. However, with the standard SiC_f/SiC samples, the creep strains to failure were low (~0.002) at all stress levels. Detailed analysis and matrix crack development, bridging the weak fiber/matrix interfaces that allow the crack faces to be bridged by intact fibers and all experimental evaluation are broadly discussed in Wilshire and Burt (2015). And these data are not reproduced

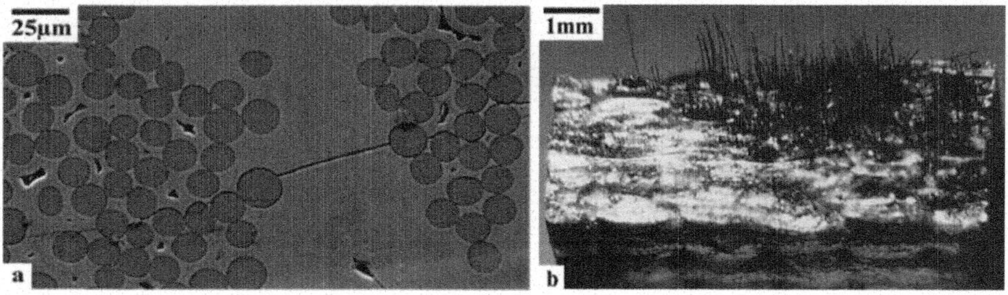

Fig. 7.2 Micrographs showing **a** matrix cracking and **b** the planar crack growth zone and final fiber pull-out region on the fracture surface of a standard SiCf/SiC sample. B. Wilshire and H. Burt, Materials Science, 2 April 30 (2013)

here because our concern is the oxygen effect. However, the bridging fibers fail progressively as oxygen penetrates into the test pieces during creep exposure. As a result, the fracture surfaces of broken specimens show planar crack growth zones, characterized by in-plane oxidation-assisted fiber failure, together with regions where sudden failure occurs by fiber pull-out (see Fig. 7.2b). Yet, while this sequence of events is applicable to the four CMCs described in Table 7.1 of Wilshire and Burt (Wilshire and Burt 2015), differences in the susceptibility of the fibers to oxidation-assisted failure affect the observed creep ductility values. With the standard SiC_f/SiC samples, cracks nucleate at macropores present in the matrix regions between the interwoven fiber bundles. These surface-nucleated cracks link up and grow, with oxygen penetrating directly along the opening crack. Rapid oxidation-assisted fiber failure

then results in high crack growth rates, with low-ductility failure occurring when the cracks reach the size required to cause sudden failure by fiber pull out (Fig. 7.2b). In contrast, with the enhanced SiC_f/SiC and $HNSiC_f$/SiC materials, the glass forming boron-based particles in the 'enhanced' matrices reduce the rates of oxygen penetration, oxidation assisted fiber failure and crack growth. Consequently, creep must continue for longer times to reach higher strains (Fig. 7.2) before the cracks attain the size needed for failure by fiber pull out.

In summary, the creep strengths of the reinforcing fibers exceed those of the matrices in SiC_f/SiC type materials. The rates of creep strain accumulation are then controlled by the longitudinal fibers, with creep of the fibers accompanied by cracking of the weak porous matrices (Fig. 7.2a). However, the creep strengths of individual fibers vary. Since the weakest fiber regions deform most easily, the creep rate decays with time as the stress is transferred to stronger fibers. The rates of crack growth are also governed by the creep resistance of the longitudinal fibers which bridge the cracks developing through the longitudinal fiber bundles. Even so, oxygen penetration into the test pieces during creep exposure promotes oxidation-assisted failure of the crack-bridging fibers, affecting the creep strains at which fracture finally occurs by fiber pull out (Fig. 7.2b). Fracture then terminates the decaying creep curves (Wilshire and Burt 2015) with the creep ductility values influencing the minimum creep rates and creep rupture lives, but also depends on the rates at which oxygen fiber/matrix interfaces are oxidized.

References

Z. Tobin, K. Petroski, P. Kerns, A. Almansour, J. Grady, S.L. Suib. ACS Omega **6**, 863 (2021)
B. Wilshire, H. Burt. Mater. Sci. **2** (2015)

Chapter 8
Fatigue-Cyclic Deformation. Effect of Environment

8.1 The Effect of Oxygen

8.1.1 In SiC

Here instead of monolithic SiC, SiC/SiC is considered, separately evaluating the monolithic crystalline SiC material, SiC fibers, and the interphase layer.

Oxidation of SiC occurs depending on if the supply of oxygen is plentiful or if it is of limited supply. In the former case a passive, protecting silica (SiO_2) layer is formed known as passive oxidation Contescu, (n d), however with limited amount of oxygen, severe corrosion occurs through oxidative volatilization and release of SiO gas phase (active oxidation).

Massive SiC oxidized in air at high temperatures produces a thin, adherent surface silica layer through the following reactions:

$$SiC_{(s)} + 3/2\ O_{2(g)} = SiO_{2(g)} + CO_{(g)} \tag{8.1}$$

$$SiC_{(s)} + 2\ O_{2(g)} = SiO_{2(s)} + CO_{2(g)} \tag{8.2}$$

The growth rate in terms of the silica layer, in terms of the scale thickness, x, or mass gain follows a linear-parabolic law with two rate constants, K_P and K_L, for the parabolic and linear regime:

$$K_P/x^2 + K_L/x = t + t \tag{8.3}$$

Here, τ is a constant that accounts for the initial scale thickness. Formation of the passive SiO_2 layer explains the outstanding oxidation resistance of silicon-based ceramic. Oxygen transfer (diffuson) through the scale depends on temperature and pressure, however below 1400 °C it is independent on oxygen pressure. At higher

J. Pelleg, *The Influence of Environment on Mechanical Characteristics of Key Semiconductor Components*, https://doi.org/10.1007/978-3-031-34530-2_8

temperatures though, permeation rates of oxygen are proportional to pressure. In an environment with low oxidizing potential, below a critical O_2 pressure (P), silica is no longer stable and SiC converts to gaseous products accompanied with weight loss according to the reaction

$$SiC_{(s)} + O_2 = SiO_{(g)} + CO_{(g)} \tag{8.4}$$

As indicated above the behavior of SiC/SiC composite is considered separately. The SiC matrix deposit is formed by chemical vapor impregnation (CVI) under pure oxygen (by passive oxidation) in the temperature range of 900–1500 °C. Except above 1500 °C the protective silica layer is amorphous.

Figures 8.1 and 8.2 show selected SEM images of oxidized surfaces of sintered α- and crystalline β-SiC of the matrix exposed to 0.2 and 100 Pa oxygen in helium. The temperatures where active to passive oxidation occurs vary, depending on the crystalline form and the oxygen pressure.

The oxidation of the SiC Nicalon fibers is active at 100 Pa O_2 and transitions to passive oxidation at 250 Pa. At higher PO_2 oxidation is passive, with mass gain caused by the growth of silica layer. The silica layer improved the high temperature behavior.

Unlimited distribution.

In summary, information regarding oxidation of SiC/SiC at high temperatures indicates improvement in the composite oxidation and its mechanical properties. The chemical stability of SiC/SiC composites in contact with oxygen or steam at temperatures that may develop in off-normal reactor conditions supports the conclusion that most advanced composites (also known as nuclear grade SiC/SiC composites) have the chemical resistance that would allow them to maintain mechanical properties at temperatures up to 1200 - 1300 °C in the extreme conditions of an air or water ingress accident scenario.

Unlimited distribution.

Continuous fiber reinforced silicon carbide ceramic matrix composites have the advantages of high temperature resistance, high strength and stiffness and low density. Because of fiber-reinforcement, the SiC/SiC composites are more damage tolerant and have the capability for larger components than their SiC monolithic counterparts and is expected to replace high-temperature alloys and refractory metals as high-temperature structural materials. Ceramic matrix composites are widely used in high temperature structural components such as heat exchangers in aerospace applications, aerospace engine nozzles, and rocket engine thrust chambers. These high temperature structural components are mostly subjected to high temperature cyclic loading during service. For the thermal structural components of aeroengines, the composite is subjected to an oxidizing gas at a temperature of 1200 C, and the internal temperature of the engine is as high as 1450 C or more. Based on the requirements of high temperature performance and overall weight reduction, continuous silicon carbide fiber toughened silicon carbide ceramic matrix composite (SiC/SiC) has become one of the main candidates for the new generation of aerospace engine thermal structural

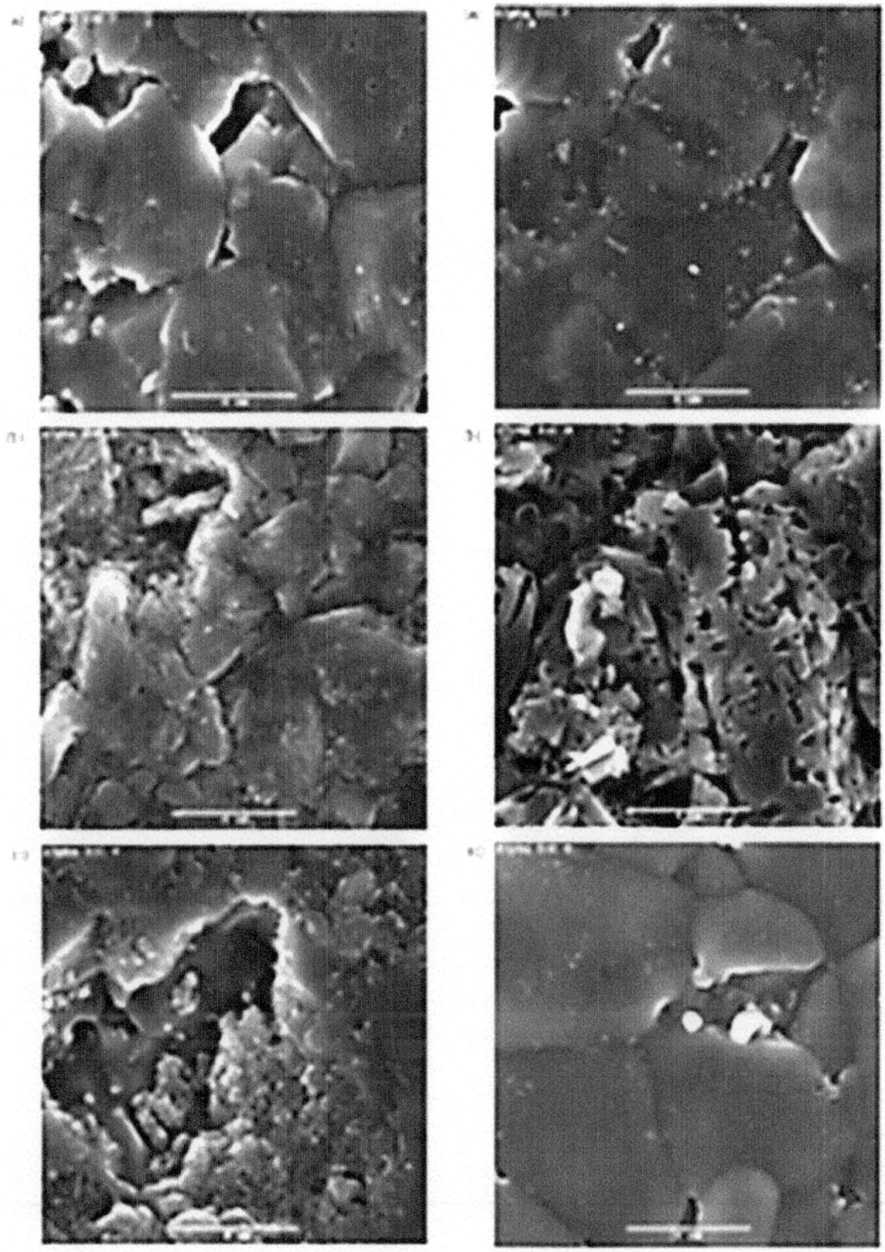

Fig. 8.1 SEM images of α-SiC oxidized under $PO_2 = 0.2$ Pa at (from top to bottom) 1553 K; 1600 K, and 1900 K (left); and under $PO_2 = 100$ Pa at (from top to bottom) 1673 K; 1762 K, and 2059 K (right). The length of scale bar is 4 μm. C. I. Contescu, OAK RIDGE NATIONAL LABORATORY Oak Ridge, Tennessee 37,831–6283

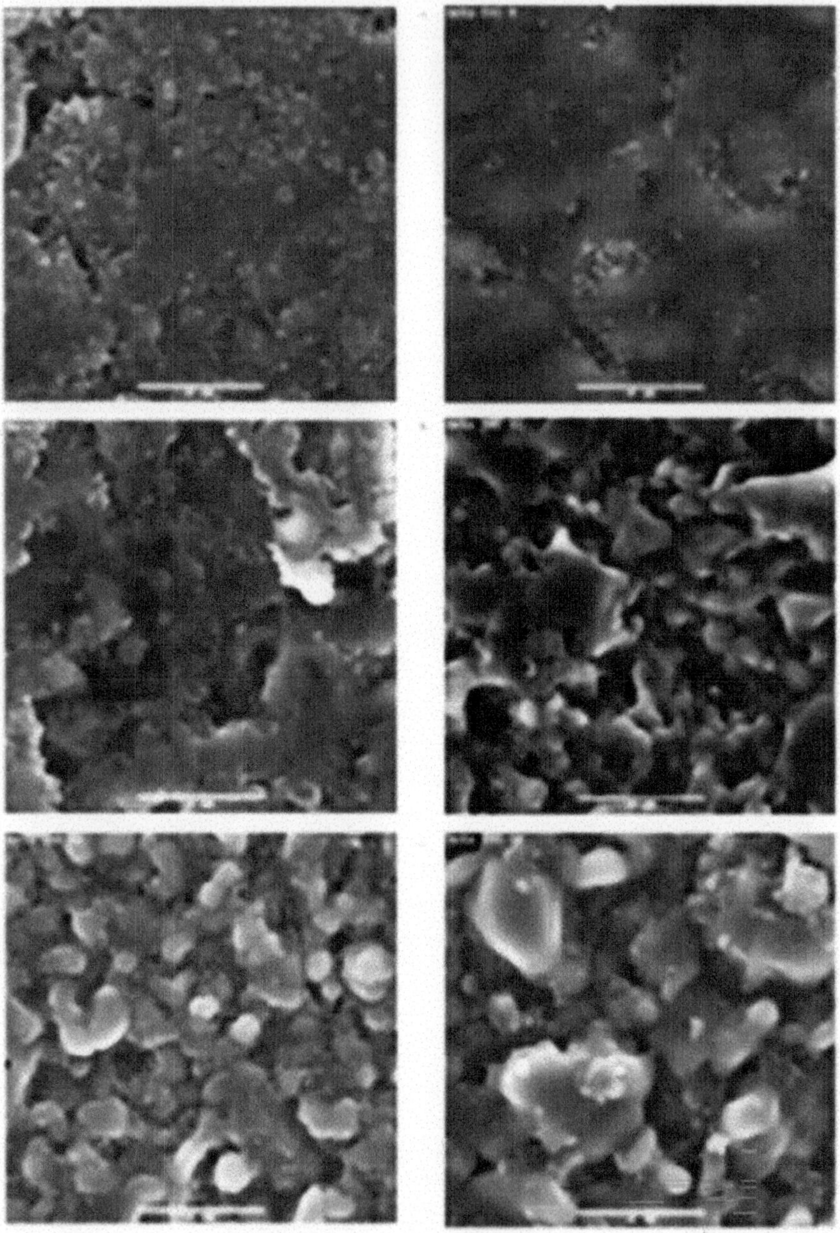

Fig. 8.2 SEM images of β-SiC oxidized under $PO_2 = 0.2$ Pa at (from top to bottom) 1595 K; 1653 K, and 2120 K (left); and under $PO_2 = 100$ Pa at (from top to bottom) 1739 K; 1900 K, and 2119 K (right). The length of scale bar is 4 μm. C. I. Contescu, OAK RIDGE NATIONAL LABORATORY Oak Ridge, Tennessee 37,831–6283

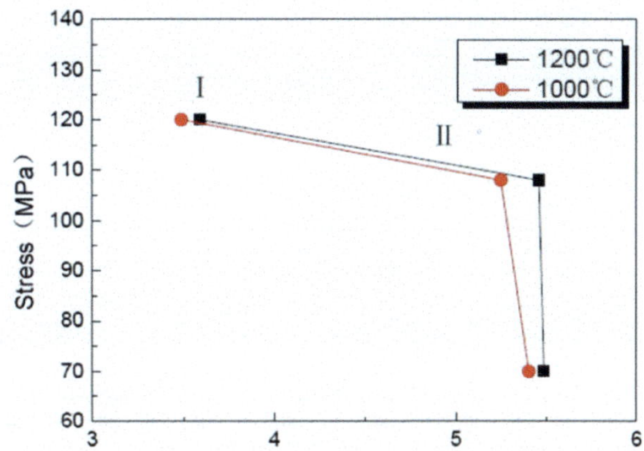

Fig. 8.3 Figure stress-life curves of 2.5D-SiC/SiC composites at 1000°C and 1200°C. Bing-yu Zhang, Hai-peng Qiu, Y. Tian, Ming-wei Chen, Shan-hua Chen, *Mater. Sci. Eng.* 678, 012,057 (2019). Open access

components with its good overall performance. The properties of SiC/SiC composites decrease with the increase of fatigue cycles. However, various damages inside the material during fatigue process among them matrix cracking and fiber breakage and interface debonding may occur or actually set in.

SiC/SiC composites were prepared via Precursor Infiltration and Pyrolysis (PIP) method (until a constant weight was achieved) and subjected to tension-tension fatigue tests. It has been shown that the fatigue life of SiC/SiC can be divided into three parts as seen in Fig. 8.3, where the peak stress of fatigue test was determined to be 35%, 55% and 60% of the tensile strength, 70 MPa, 108 MPa and 120 MPa respectively.

Figures 8.4, 8.5 and 8.6 are SEM micrographs of fatigue specimens tested at the conditions indicated in the subscripts. The test and the fracture occur in air environment. The sample shows some ductile fracture. Toughening mechanisms such as crack deflection and fiber bridge chain can be seen in SiC matrix, indicating that the failure mechanism of materials under tensile fatigue load of 1200°C and 70 MPa is similar to that of uniaxial tensile strength.

In summary, the main reason that causes tension-tension fatigue failure of SiC/SiC composites are oxidation damage and stress damage. The microstructure of SiC/SiC composites demonstrated that the oxidation of interfaces and SiC fibers caused the failure of the composites. Under the load of alternating stress, the original voids and cracks in the material preparation process will be widened, and new cracks will be generated, for cracked SiC/SiC composites. Oxygen is especially serious at intermediate temperatures(~800°C) where it can reach the SiC fiber before being sealed off by slow-growing silica on the matrix.

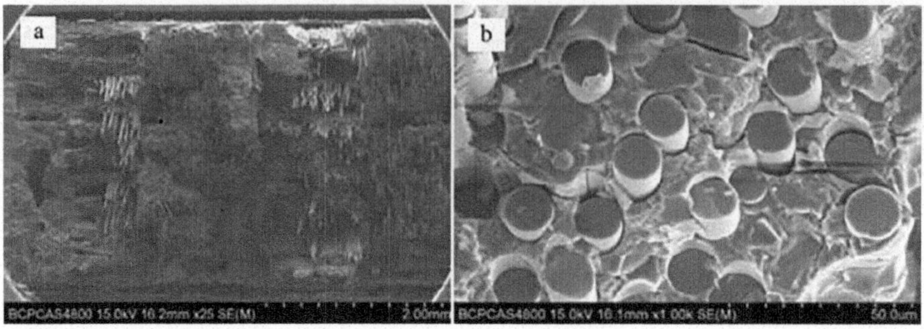

Fig. 8.4 SEM micrographs of specimen tested in fatigue at 1Hz in air at 1200 C ($\sigma_{max} = 70$ MPa, $N_f = 305{,}984$, R = 0.1). Bing-yu Zhang, Hai-peng Qiu,Y. Tian, Ming-wei Chen, Shan-hua Chen, *Mater. Sci. Eng.* 678, 012,057 (2019). Open access

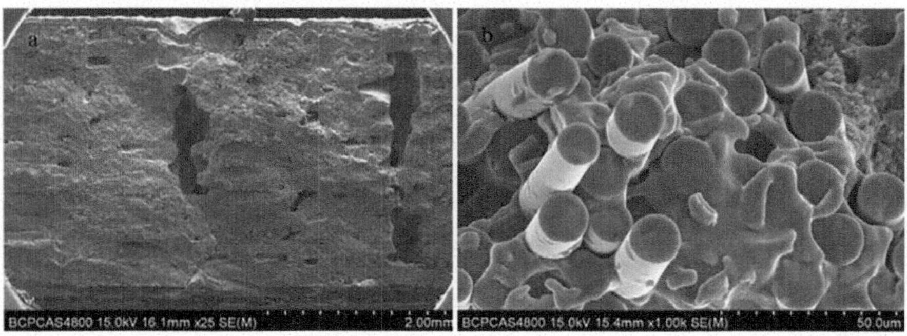

Fig. 8.5 SEM micrographs of specimen tested in fatigue at 1Hz in air at 1000 °C ($\sigma_{max} = 70$MPa, $N_f = 28{,}801$, R = 0.1). Bing-yu Zhang, Hai-peng Qiu,Y. Tian, Ming-wei Chen, Shan-hua Chen, *Mater. Sci. Eng.* 678, 012,057 (2019). Open access

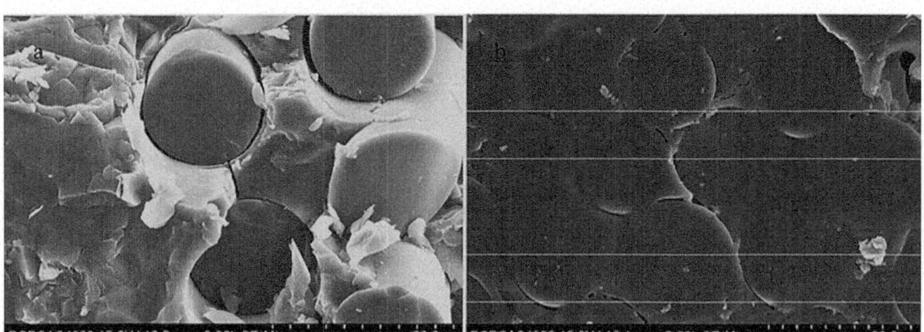

Fig. 8.6 SEM micrographs of specimen tested in fatigue at 1Hz in air at **a** 1200 °C ($\sigma_{max} = 108$MPa, $N_f = 252{,}988$, R = 0.1) and **b** 1000° C ($\sigma_{max} = 108$MPa, $N_f = 177{,}010$, R = 0.1). Bing-yu Zhang, Hai-peng Qiu, Y. Tian, Ming-wei Chen, Shan-hua Chen, *Mater. Sci. Eng.* 678, 012,057 (2019). Open access

8.2 The Effect of Moisture

8.2.1 *In SiC*

An important ceramic matrix composite (CMC) at elevated temperature use is silicon.

carbide fiber reinforced in silicon carbide matrix (SiC) with boron nitride (BN) interphase. In this section tension-tension fatigue tests performed at 750 °C at a frequency of 10 Hz with a stress ratio of 0.1 is described. The tests were conducted at (1) under dry environment test condition with 0% moisture content and (2) under humid environment test condition with 60% moisture content. Moisture decreased the fatigue life at a prescribed stress level relative to that without moisture. A detailed microscopic analysis showed that moisture caused embrittlement of the interphase between fiber and matrix, which in turn caused the degradation of fatigue performance of the tested CMC system.

Note, that high strength ceramics are appealing for applications because they possess high strength and modulus of elasticity at elevated temperatures, but at the same time -unfortunately- the exceptional strength and stiffness also causes embrittlement at a very intolerable level. Thus, monolithic ceramics-including SiC- fail catastrophically and without warning. With the addition of high strength reinforcing fibers/ fabrics however, the overall strength of the ceramic, as well as its fracture toughness, can be increased significantly. CMCs are the leading candidates for a number of applications in turbine engines for aircraft and future space vehicles (reusable launch vehicles, space operation vehicles etc.), land-based power generation systems, and automobiles.

The available SiC fibers in use with SiC matrix are Nicalon (Nic), Hi Nicalon and Sylramic (Syl). The interest in C as interphase has decreased because its poor performance in oxidizing atmosphere (also volatizing), thus BN remains as the interested interface. The system to be described below is Syl-iBN/BN/SiC which is composed of Syl fibers, BN interphase and SiC matrix and is one of the strongest CMC and the most creep resistant system. The system preparation and all the experimental steps are presented in Zhang et al., 2019.

The applied stress to failure in the Si/SiC composite in two test environments of 0% (dry) and 60% moisture is presented. To receive information prior to the fatigue tests monotonic tensile stress–strain tests were performed, to which the applied cyclic stresses were normal1zed by the average monotonic strength at 750 °C being 345 MPa (See Fig. 2 of the MS in Zhang et al., 2019). The normalized peak stress in the two conditions is seen in Fig. 8.7. The variation of maximum and minimum strains during a fatigue test is shown in Fig. 8.8, showing the environment test of 0% and 60% moisture. Further, it appears that (from monitoring of strain) that more damage is present in humid environment having 60% moisture content than that in the dry condition having 0% moisture content. This is seen in Fig. 8.9 where the Young's modulus from these tests are compared.

The values of modulus in the figure are near the failure point of fatigue. The figure shows that specimen had less stiffness at a given cyclic stress level at or near failure

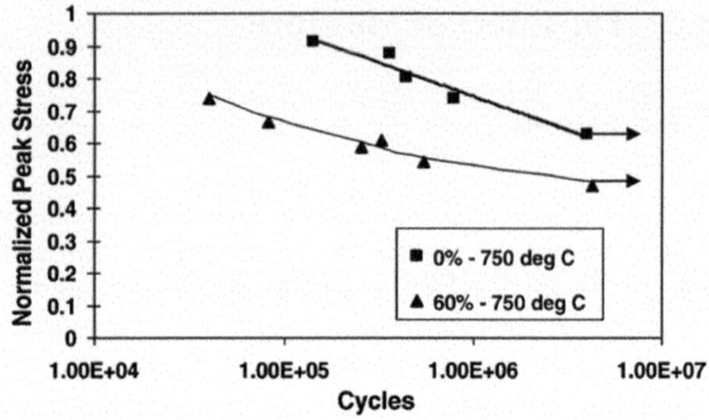

Fig. 8.7 Applied stress vs. cycles to failure relationships. Shankar Mall, Mater. Sci. Eng. A 412, 165 (2005). With kind permission of Elsevier

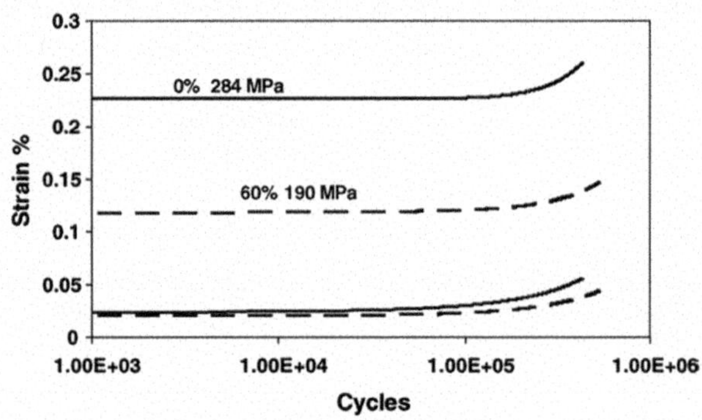

Fig. 8.8 Typical evolution of strain during fatigue test (upper and lower curves represent maximum and minimum strains). Shankar Mall, Mater. Sci. Eng. A 412, 165 (2005). With kind permission of Elsevier

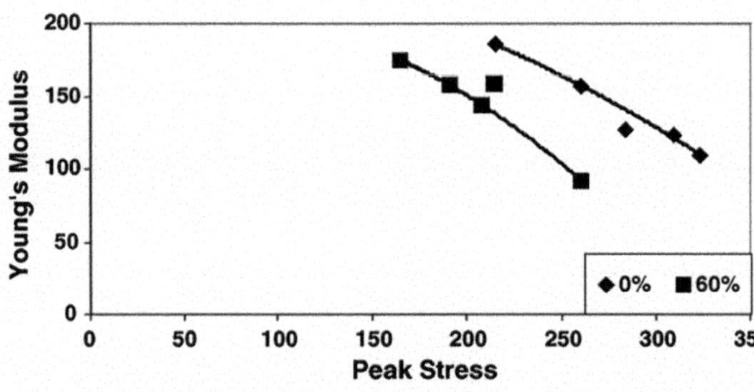

Fig. 8.9 Young's modulus vs. applied stress at the end of fatigue test. Shankar Mall, Mater. Sci. Eng. A 412, 165 (2005). With kind permission of Elsevier

under the humid test condition than that under dry test condition. A comparison between monotonic fracture surface and those of fatigue tests under dry and humid conditions are shown in Figs. 8.10, 8.11 and 8.12, respectively. Figures 8.13 and 8.14 compare the fracture surfaces of fatigue tests from both test condition at a higher magnification which clearly shows the embrittlement of the fiber/matrix interphase in

Fig. 8.10 Close-up of fracture surface (monotonic test). Shankar Mall¸ Mater. Sci. Eng. A 412, 165 (2005). With kind permission of Elsevier

the presence of the moisture (Fig. 8.14). The BN interphase between fiber and matrix showed evidence of brittle fracture whereas no degradation in the dry test condition as seen in Fig. 8.13. The BN interphase appears to remain intact at most of the location. The BN interphase near the fiber/matrix or fiber/fiber locations appears to have been oxidized by the moisture under humid test environment, which occurred most probably due the oxidation of BN to form boria (B_2O_3). It ultimately after various reactions causes an embrittlement. The experimental data clearly suggests that the tested composite experienced the embrittlement of fiber/matrix interphase in the presence of moisture, which in turn caused a reduction in fatigue performance relative to that in the dry (no moisture) condition.

In summary, tension–tension fatigue tests were conducted at 10 Hz with a stress ratio of 0.1 at 750 °C: (1) under a dry environment test condition (with 0% moisture content) and (2) under a humid environment test condition (with 60% moisture content). Fatigue life at a prescribed stress level was shorter in presence of moisture than that without moisture. Microscopic analysis showed that presence of moisture in the test environment caused the embrittlement of the BN interphase between fiber/ matrix. The BN oxidized to form boria which then caused a reduction in the fatigue performance relative to that in the dry (no moisture) condition.

Fig. 8.11 Close-up of fracture surface (fatigue test under dry condition). Shankar Mall¸ Mater. Sci. Eng. A 412, 165 (2005). With kind permission of Elsevier

Fig. 8.12 Close-up of fracture surface (fatigue test under humid condition). Shankar Mall¸ Mater. Sci. Eng. A 412, 165 (2005). With kind permission of Elsevier

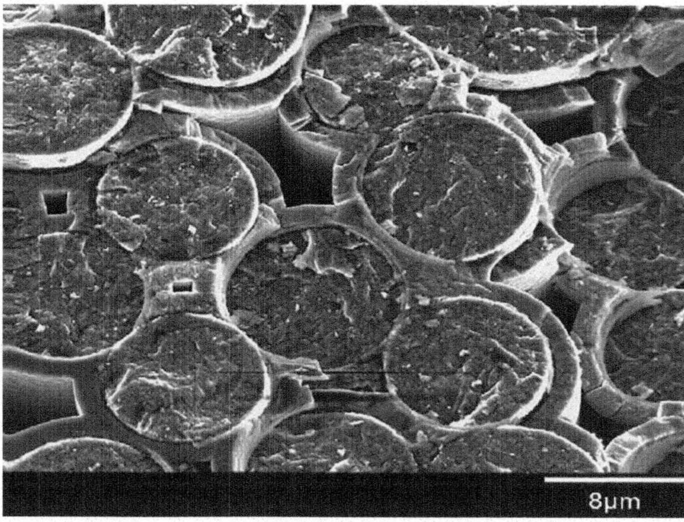

Fig. 8.13 Close-up of fiber/matrix interphase from fatigue test under dry condition; maximum stress = 284 MPa, cycles to failure = 4.35E06, 750 °C. Shankar Mall¸ Mater. Sci. Eng. A 412, 165 (2005). With kind permission of Elsevier

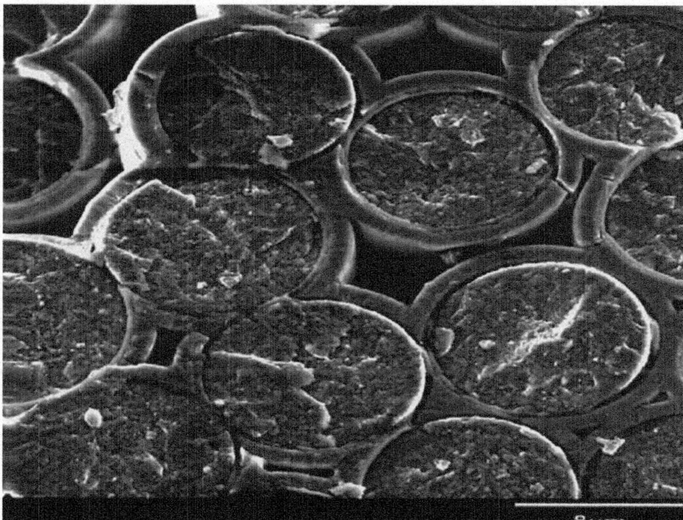

Fig. 8.14 Close-up of fiber/matrix interphase from fatigue test under humid condition; maximum stress = 165 MPa, cycles to failure = 4.25E06, 750 °C. Shankar Mall¸ Mater. Sci. Eng. A 412, 165 (2005). With kind permission of Elsevier

References

C. I. Contescu, Oak ridge national laboratory oak ridge, Tennessee 37831–6283
S. Mall, Mater. Sci. Eng. A 412, 165 (2005)
B. Zhang, H. peng Qiu,Y. Tian, M. Chen, S. Chen, Mater. Sci. Eng. 678, 012057 (2019)

Chapter 9
Fracture—Effect of Environment

9.1 The Effect of Oxygen

9.1.1 In Si

Low-oxygen concentration in silicon wafers is presented in this section considering slip by using three-point bend test. Use of wafers in insulated gate bipolar transistors (IGBTs) require oxygen precipitate free wafers. However, in floating zone (FZ) silicon wafers, which are often used in IGBT, there are a few oxygen atoms in the bulk of the wafer, in which the motion of slip dislocations is not hindered. If, however in the IGBT manufacturing process scratches are introduced, they become the starting point of slip, and slip dislocations may easily move under thermal treatment process. The migration of slip dislocations has been known to cause current leakage and overlay issues in the photolithography process, which becomes significantly problematic during device production.

Starting points of slip dislocations are either (a) due to oxygen precipitates and, (b) the back surface of the wafer. Regarding (a) there is no concern over slip generated from oxygen precipitates in low-oxygen wafers; regarding (b) the wafer comes into contact with various materials (containers during transportation, etc.) and thus contact scratches are always introduced onto the back surface of the wafers. Consequently, slip is generated leading to deformation and current leakage.

Three-point bending test at high temperature is used to introduce scratches to evaluate the critical shear stress for slip generation. The critical shear stress is dependent on temperature, oxygen concentration, and strain rate. Experiments with FZ crystals have been conducted to clarify the yield stress of silicon and the dependence of the dislocation velocity on the oxygen concentration and temperature, understanding of which is important for improving IGBT manufacturing process, using low oxygen wafers. Furthermore, the critical stress increases when the oxygen concentration below the surface was increased through thermal treatment. When lowoxygen wafers such as IGBTs are used, appropriate thermal treatment can prevent slips generation

J. Pelleg, *The Influence of Environment on Mechanical Characteristics of Key Semiconductor Components*, https://doi.org/10.1007/978-3-031-34530-2_9

from scratches, thereby improving the production process. In the experiments, first the oxygen concentration below the surface of the wafer is changed by preannealing, then by Vickers indentation indents (0.8 N) were performed to generate dislocations around the indent and increasing the temperature and then stress was applied using a high temperature three-point bending test. The atmosphere was the same as in air controlled clean room. The three-point bending test set-up is shown schematically in Fig. 9.1. As seen the shear stress decreases from the center to the outside, so that the slip dislocation does not extend in the area where the shear stress is less than critical shear stress. The critical shear stresses of each specimen were calculated using the critical distance X and the detailed analysis and experimental procedure are given in Fujise et al. 2020. The critical shear stress is shown in Fig. 9.2 at 1023 K. The critical distance, X did not change with increased time, and the dwell time of the stress was not affected from the extension of the slip from the indent by the shear stress. The slip length L_s can be calculated from the dislocation velocity equation given as stress applied, k (8.62×10^{-5} eV K^{-1}) is the Boltzmann constant, and T is the absolute temperature. The critical shear stresses of each specimen are shown in Fig. 9.3 at 1123 K. The critical shear stress obtained from the X-ray image is shown in Fig. 9.4. It can be seen that the higher the thermal treatment temperature and the longer the time, the higher the critical shear stress. The critical shear stress is known to be affected by the oxygen through locking effect. The locking stress can be expressed by the interstitial oxygen through locking effect. The locking stress can be expressed as

$$Ls = vt = 1 \times 10^{-2} \tau \exp\left(-\frac{2.2ev}{KT}\right) t \tag{9.1}$$

where v is the dislocation velocity, t is the time of stress applied, τ is the shear

$$\tau_u = AC_0 \exp\left(-\frac{E_a}{KT}\right) \tag{9.2}$$

where A is temperature independent prefactor, C_o (10^{17} atoms cm^{-3}) is the oxygen concentration and E_a is the activation energy for releasing a dislocation from locking oxygen atoms. The values of A and E_a were obtained by fitting τ_u curves to the experimental data, then Eq. 9.3 is obtained given as

$$\tau_u = 5.3 \times 10^{-5}\, C_0 \exp\left(-\frac{0.88ev}{KT}\right) \tag{9.3}$$

In this experiment, a critical shear stress of approximately 1.3 MPa was detected regardless of the temperature. This suggested the existence of some long-range obstacles which causes athermal stresses even in single crystalline silicon.

It was observed as can be seen (Fig. 9.4) that the higher the pre-annealing temperature and the longer the pre-annealing time, the higher the critical shear stress. This can probably be attributed to the change in the oxygen concentration below the

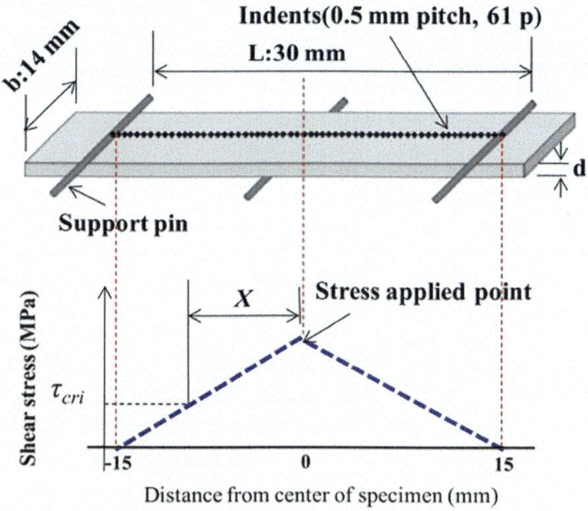

Fig. 9.1 Schematic representation of the three-point bending test. Specimen width b is 14 mm and the distance between support pins L is 30 mm. Indents introduced by Vickers hardness tester on the specimen surface are placed at 0.5 mm pitch. The graph included below represents the profile of shear stress as a function of the distance from the center of the specimen. J. Fujise, B. Ko, T. Ono and M. Tanaka, ECS J. Solid State Sci. Tech., 9, 104002 (2020). Open access

Fig. 9.2 Critical shear stress as a function of stress dwell time at test temperature (1023 K). J. Fujise, B. Ko, T. Ono and M. Tanaka, ECS J. Solid State Sci. Tech., 9, 104002 (2020). Open access

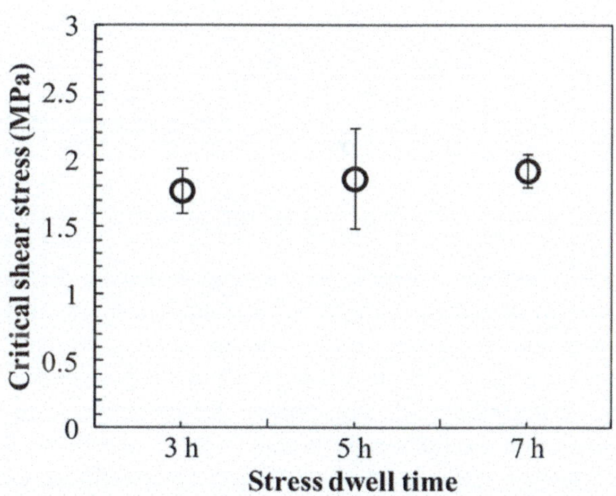

surface of the wafer. The oxygen concentration was measured by SIMS. Figure 9.5 shows the depth profiles of oxygen concentration from wafer surface region after preannealing.

Table 9.1, indicated in Fig. 9.5 is a summary of the experimental results. But beforehand Figs. 9.6 and 9.7 are illustrated. Figure 9.6 is a TEM image showing the dislocation density under the indent of 0.98 N. At the depth 3.3 to 7.3 mm is the slip

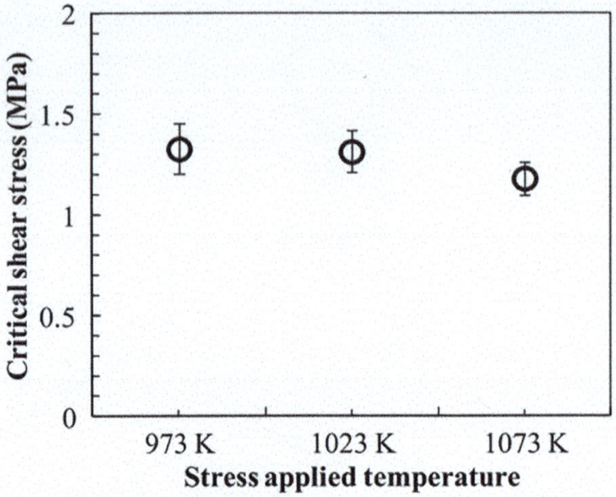

Fig. 9.3 Critical shear stress as a function of test temperature. Shear stress of 2.9 MPa was applied for 7 h at 973 K and 1023 K and then for 1 h at 1123 K. J. Fujise, B. Ko, T. Ono and M. Tanaka, ECS J. Solid State Sci. Tech., 9, 104002 (2020). Open access

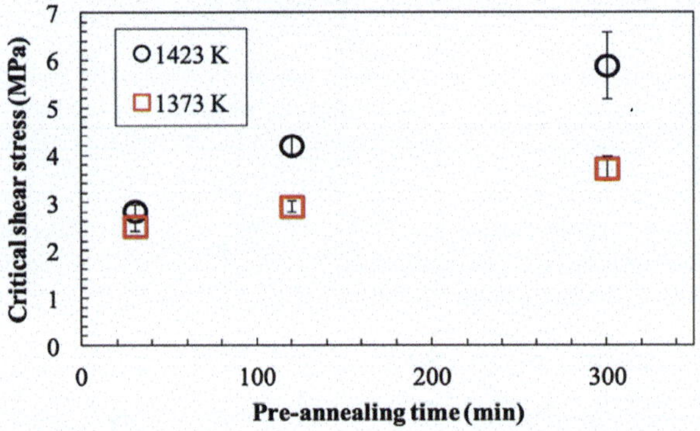

Fig. 9.4 Changes in critical shear stress at various pre-annealing conditions. J. Fujise, B. Ko, T. Ono and M. Tanaka, ECS J. Solid State Sci. Tech., 9, 104002 (2020). Open access

generation depth where the dislocation density decreases with the depth from the surface. In general, at high dislocation density, the dislocations interfered with each other, and their motion require higher stresses, which might explain why slip was not generated at 3.3 mm or below it, where the dislocation density is high since it is just below the indent. Since the loads of all indents were the same 0.98 N, the depth of slip generation should also be the same (depth variation 3.3–7.3 mm). The average depth of slip generation is 5.6 mm. There is a proportional relationship between the oxygen concentration and the critical shear stress at the same depth (5.6 μm). A specimen with a constant oxygen concentration in the depth was used to calculate the critical shear stress using Eq. 9.3. It is expected that the failure during the device process can be reduced by paying attention to the oxygen concentration distribution in the depth direction. For example, in a low-oxygen concentration wafer, it is possible to prevent slip generation during the device process by performing pre-annealing over a long time at a high temperature to increase the oxygen concentration below the wafer surface.

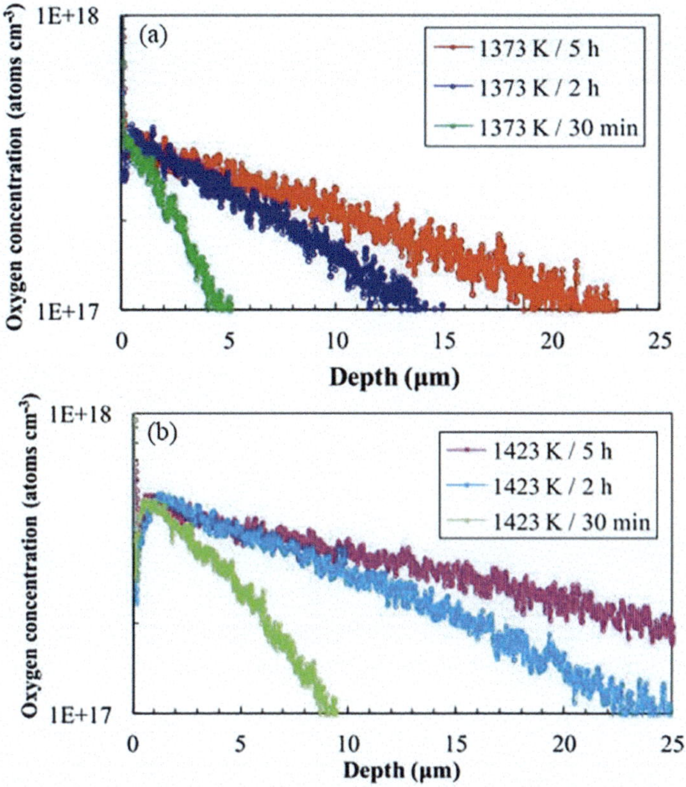

Fig. 9.5 Depth profiles of oxygen concentration from wafer surface region after each pre-annealing condition described in Table I. **a** 1373 K, and **b** 1423 K. J. Fujise, B. Ko, T. Ono and M. Tanaka, ECS J. Solid State Sci. Tech., 9, 104002 (2020). Open access

Wafers with low oxygen concentrations (including FZ wafers) are used for IGBT manufacturing. Such wafers with low oxygen concentrations have low locking stress, so slip dislocation can easily be extended. However, it was confirmed that the dislocation motion could be suppressed by increasing the oxygen concentration below the surface by pre-annealing. By using this result, it is possible to reduce the risk of wafer bending and leakage current due to slip during manufacturing. Table 9.1 summarizes the results of this work.

In summary, a high temperature bending test was performed using a low oxygen FZ silicon wafer for IGBT. The temperature dependence of the critical stress and the oxygen concentration distribution along the depth direction from the wafer surface were obtained. The results showed that the critical stress of FZ wafers had no temperature dependence above 973 K. It was also found that the oxygen concentration distribution along the depth direction from the wafer surface had a significant influence on the critical stress. Controlling the oxygen concentration by thermal treatment

Table 9.1 A summary of results. J. Fujise, B. Ko, T. Ono and M. Tanaka, ECS J. Solid State Sci. Tech., 9, 104002 (2020). Open access

Summary of critical shear stress, oxygen concentration and depth of slip generation at each pre-annealing condition

Pre-annealing temperature	Pre-annealing time	Critical shear stress	Calculated Co by Eq. 6	Depth of slip generation
1373 K	30 min	2.5 MPa	1.7×10^{17} atoms cm^{-3}	3.3 μm
1373 K	2 h	2.9 MPa	2.0×10^{17} atoms cm^{-3}	5.0 μm
1373 K	5 h	3.7 MPa	2.5×10^{17} atoms cm^{-3}	6.1 μm
1423 K	30 min	2.8 MPa	1.9×10^{17} atoms cm^{-3}	5.0 μm
1423 K	2 h	4.2 MPa	2.9×10^{17} atoms cm^{-3}	7.3 μm
1423 K	5 h	5.9 MPa	4.0×10^{17} atoms cm^{-3}	6.7 μm

suppresses slip generation during the device process, which can solve the problem of wafer bending and current leakage during manufacturing.

9.1.2 *In SiC*

Stress rupture of SiC/SiC composites at intermediate temperatures in oxidizing environments is the result of a series of internal chemical and thermomechanical processes that lead to premature, localized fiber fracture.

A relatively novel approach is to use as the reinforcing agent woven SiC fabric with SiC matrix comprising the composite. The fracture behavior and microstructure of two dimensional woven SiC fibers woven together to form a cloth as the reinforcing agent was exposed to surface temperature over 1400 °C to evaluate its functional performance. It is of major importance to understand the degradation rate of the Nicalon SiC fiber (that forms the fabric) in oxygen environment and its propensity of embrittlement. Hence the fracture strength reduction and embrittlement in 2D woven SiC, i.e., Hi-Nicalon SiC in oxygen plasma (producing high surface temperature) is presented below. High-temperature aero-thermal heating in a 30 kW inductively coupled plasma torch was used to replicate the effects of harsh oxidizing environments during hypersonic atmospheric entry on fracture behavior and microstructure of two-dimensional woven SiC fibers.

SiC tows were woven together to form a cloth, which are untwisted bundles of parallel, continuous filaments as shown in Fig. 9.8 a. They were woven together

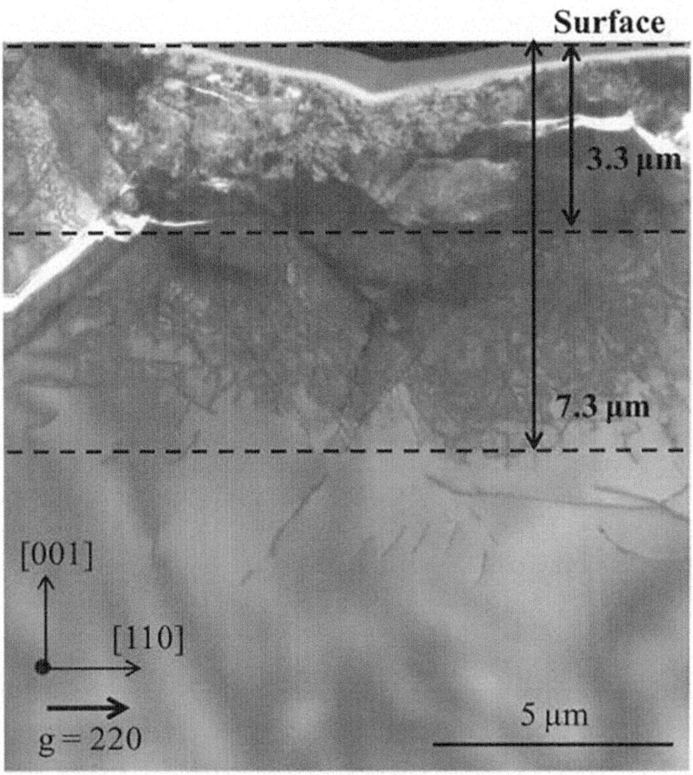

Fig. 9.6 Dark field transmission electron microscopy image displaying the crosssection just beneath the indents (0.98 N) under a two-beam condition. The diffraction vector was set to be g(hkl) = 220. The incident beam direction is [110]. In the image, the range of slip generation depth is 3.3 − 7.3 μm, and the dislocation density decreases as the depth increases. J. Fujise, B. Ko, T. Ono and M. Tanaka, ECS J. Solid State Sci. Tech., 9, 104002 (2020). Open access

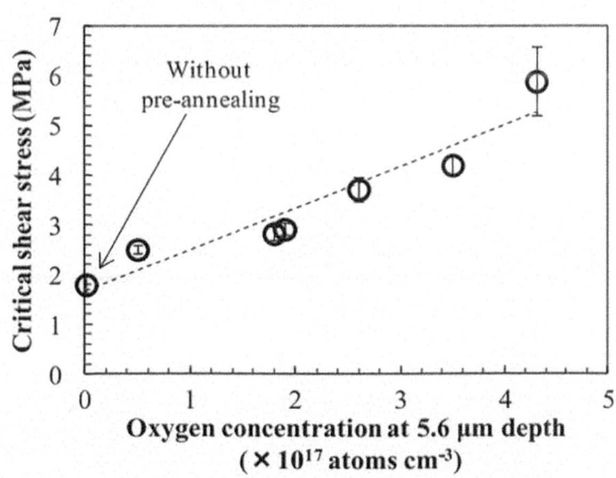

Fig. 9.7 Relationship between critical shear stress and oxygen concentration at a depth of 5.6 μm from the specimen surface. J. Fujise, B. Ko, T. Ono and M. Tanaka, ECS J. Solid State Sci. Tech., 9, 104002 (2020). Open access

to form a cloth, as shown Figs. 9.8b, c. Gap between the fibers allowed for gas-surface interactions to take place around the circumference of individual filaments. A profile view of the inductively couple plasma (ICP) test configuration is displayed in Figs. 9.8d, e. The Aerothermal conditions tested in these Figs. 9.8d, e included 100% nitrogen, dilute air, normal air, and 100% oxygen plasmas at a constant chamber static pressure of 160 torr and are summarized in Table 9.2. 20-mm diameter coupons were made by mechanically punching a sheet of SiC fabric. The fabric coupons had a bulk thickness of 0.5 mm and an average mass of 0.14 g. The coupons were held at a fixed position by an electrically insulated arm which was used to quickly insert the samples into an already established plasma stream. Two outer fabric coupons were tested simultaneously. similar to the lead-ply and sub-ply structure used in inflatable aerodynamic decelerators (IAD). The lead-ply coupon was exposed directly to the plasma flow while the sub ply was mildly protected between the lead ply and the SiC backing piece. The operating conditions were chosen to provide similar steady state temperatures for all gases. Figure 9.9 shows that the samples tested in pure air and airnitrogen plasmas attained relatively constant surface temperatures within ± 25 °C after 5 s. Figure 9.9a displays a representative load–displacement curve showing that the cloth deformation is quasi-linear with a clean break at peak load $F_{max.}$ The fracture strength s_f was calculated by using where f is the total number of failed fibers, and D is the average fiber diameter. f was determined by counting the total number of broken tows times the average numbers per tow. Using SEM analysis, we found D ~ 14 mm and an average of 500 fibers per tow. Optical and SEM micrographs illustrated in Fig. 9.10 show the fabric surface after exposure to oxidizing plasmas where a vitreous silica layer seems to be present. Figure 9.99..11 shows that fiber oxidation as a function of time in both air and oxygen plasmas leads to severe strength reduction and embrittlement compared to nitrogen plasmas as the time of exposure increases. Strength reduction for the sub ply occurs at an initially slower rate compared to the lead-ply coupon, as shown in Fig. 9.11b. This is a consequence of suppression of gas-surface interactions deeper within the weave. Nevertheless, total strength loss at room temperature was manifested after 240 s from insertion for both lead-ply and sub-ply coupons, which shows a fundamentally different embrittlement behavior than that of woven SiC fibers treated in pure nitrogen plasma environment.

In summary, exposure to high oxygen atom concentration in air and oxygen plasmas result in severe embrittlement over a shorter time scales than those reports for static heating in conventional furnaces where only molecular oxygen is present. The origin of the accelerated embrittlement in oxidizing plasmas was found to be associated with the formation of a viscous silica surface layer leading to the development of gas bubbles, inter-filament adhesion, and critical flaws. Further, results obtained using admixtures of oxygen with argon to assess the influence of temperature suggested that the fracture resistance of 2-D woven SiC fibers is totally lost after 120 s of plasma exposure at 1400 °C. The work of Owens et al. 2015 should be consulted for a detailed analysis of the experimental results and their interpretation.

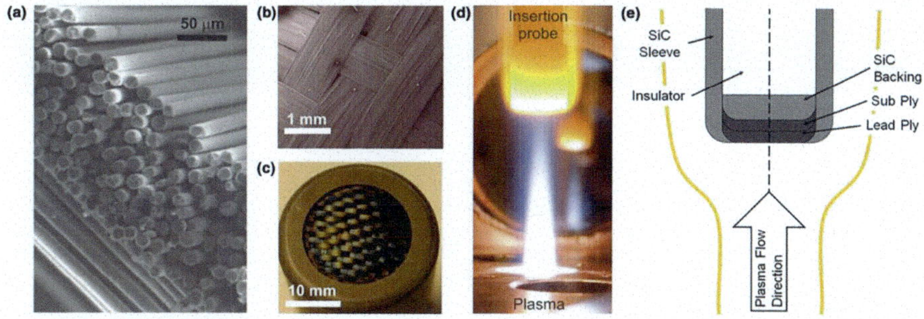

Fig. 9.8 Aerothermal testing of 2-D woven SiC fibers. **a** Cross-sectional and **b** top views of woven SiC fibers by SEM. **c** Woven Hi-Nicalon SiC lead-ply coupon in SiC sleeve. **d** Probe insertion in the 30 kW inductively coupled plasma torch facility **e** Schematic illustration of a longitudinal cross-section view of the insertion probe. W. Owens, D. Merkel, F. Sansoz, and D. Fletcher, J. Am. Ceram. Soc., 98, 4003 (2015). With kind permission of John Wiley and Sons

Table 9.2 ICP Torch operating conditions for aerothermal testing. Tests A1–A4 were Used to Characterize the Effects of Plasma Environment, and Tests B1–B4 for the Effects of Surface Temperature. Slpm is standard liter per minute. W. Owens, D. Merkel, F. Sansoz, and D. Fletcher, J. Am. Ceram. Soc., 98, 4003 (2015). With kind permission of John Wiley and Sons

Plasma conditions					Temperature (C)	
Sample	Gas mixture	Gas flow slpm, (±10%)	Power. kW (±2 kW)	Heat flux, W/ cm^2 (±15%)	Lead ply (±50"C)	Sub ply (±75 °C)
A1	Nitrogen	40	14	62	See Fig. 2	
A2	Air: Nitrogen	15: 30	14	73	See Fig. 2	
A3	Air	40	14	80	See Fig. 2	
A4	Oxygen	40	12	79	See Fig. 2	
B1	Oxygen: Argon	10: 30	3.7	26	846	726[a]
B2	Oxygen: Argon	10: 30	6.6	57	1262	1142
B3	Oxygen: Argon	10: 30	7.5	78	1420	1300
B4	Oxygen: Argon	10: 30	10.3	107	1533	1413

[a](±10°C) error

9.1.3 In GaAs

Currently, worldwide end uses for GaAs comprise integrated circuits (ICs), optoelectronic devices, including light emitting diodes (LEDs), laser diodes, photodetectors.

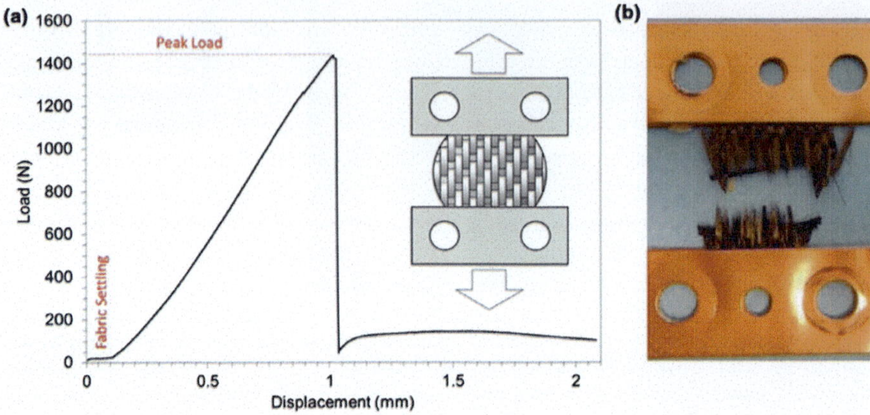

Fig. 9.9 Tensile testing of 2-D woven SiC fibers. **a** Representative loaddisplacement curve showing a clean break at peak load. Inset shows the yarn orientation with respect to the loading direction. **b** Coupon after fracture. W. Owens, D. Merkel, F. Sansoz, and D. Fletcher, J. Am. Ceram. Soc., 98, 4003 (2015). With kind permission of John Wiley and Sons

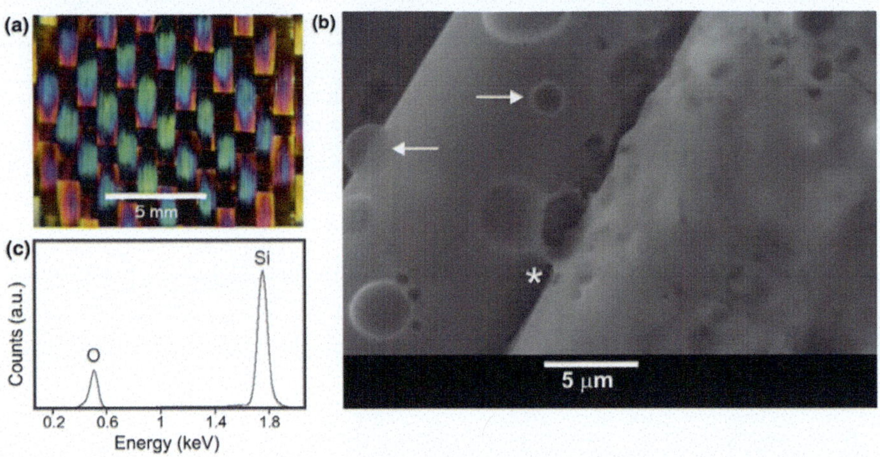

Fig. 9.10 Structure of 2-D woven SiC fabric after pure oxygen plasma exposure for 240 s. **a** Optical micrograph and **b** SEM image of the exposed lead-ply surface with arrows showing the formation of gas bubble at high temperature and a star for an example of surface bonding between two fibers. **c** EDS spectrum of the silica surface. W. Owens, D. Merkel, F. Sansoz, and D. Fletcher, J. Am. Ceram. Soc., 98, 4003 (2015). With kind permission of John Wiley and Sons

Heat and moisture resistance, and superior flexibility are the distinct advantages of GaAs, which are improving the acceptance of GaAs wafers in the semiconductor industry. Liquid encapsulation Czochralski (LEC) and vertical gradient freeze (VGF) are two popular methods which are improving the production of GaAs wafers with high uniformity of electrical properties and excellent surface quality.

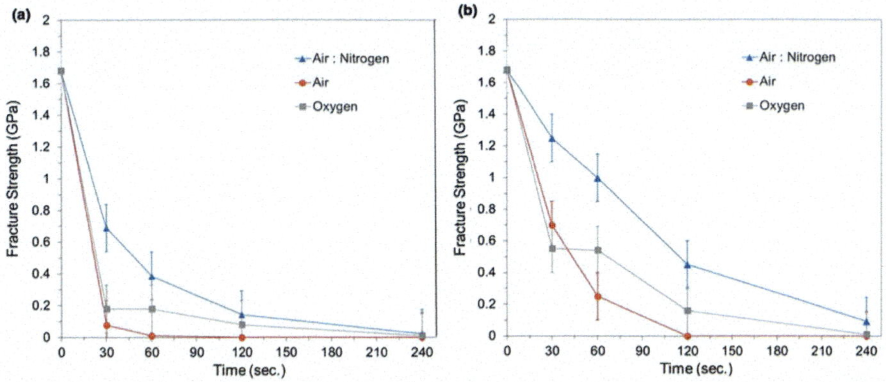

Fig. 9.11 Fracture strength after exposure to different oxidizing conditions for **a** lead-ply and **b** sub-ply coupons. W. Owens, D. Merkel, F. Sansoz, and DFletcher, J. Am. Ceram. Soc., 98, 4003 (2015). With kind permission of John Wiley and Sons

Electronic devices have a wide range of failure modes. These can be related by their development in time -sudden failure or gradual degradation- or by environmental effects, i.e. corrosion, ionizing radiation, or by the electrical parameter which was exceeded such as electrostatic discharge, overvoltage, overcurrent, etc.. Failures most commonly occur at the beginning and near the end of lifetime of the parts.

It was felt that this short section might be helpful, even if detailed data are not presented due to the lack of information and the limited data that are available in the open literature.

References

J. Fujise, B. Ko, T. Ono, M. Tanaka, ECS J. Solid State Sci. Tech. **9**, 104002 (2020)
W. Owens, D. Merkel, F. Sansoz, D. Fletcher, J. Am. Ceram. Soc. **98**, 4003 (2015)

Chapter 10
Twinning–Effect of Environment

10.1 Introduction

Twinning is an important mechanism of plastic deformation in some metallic materials. It is a shear force that can produce atomic displacements. Twinning may be produced by mechanical deformation or as the result of annealing following plastic deformation. One is pondering if nonmetallic materials can twin. It turns out that under certain specific condition also in materials such as silicon, silicon carbide or gallium arsenide-which are a subject of this book- twinning has been recorded. Thus, in this chapter the elements of twinning in these materials will be considered. But beforehand the concept of twinning will be introduced shortly.

A variety of crystalline solids possessing low stacking fault energy show the propensity to undergo twinning during solidification, deformation, solid state phase transformation or recrystallization. Thus, one can talk on mechanical twinning (due to the mentioned deformation) and annealing twinning (transformation or recrystallization involve thermal exposure and then cooling). Twinning is observed since their lower interfacial energy than those of general grain boundaries. By convention, a deformation twin in a crystal is a region where the deformation, usually shear, occurs in such a way that the resulting structural shape is identical to that of the parent matrix except for its orientation. Here it is the place to realize the difference between slip and twinning deformations. The key difference is that slip involves the movement of all atoms the same distance in a block of crystal, while in twinning the atoms in each successive plane of a block move through different distances that are proportional to their distance from the twinning plane. Twinning and slip are competitive mechanisms for crystal deformation. Each mechanism is dominant in certain crystal systems and under certain conditions. In FCC metals, slip is almost always dominant because the stress required is far less than twinning stress.

Below an accepted summary on the origin of twinning is presented Prof et al. 2019. Three ways are indicated and referred to as: (1) growth twins, (2) transformation twins and (3) deformation twins sometimes referred to as glide twins.

J. Pelleg, *The Influence of Environment on Mechanical Characteristics of Key Semiconductor Components*, https://doi.org/10.1007/978-3-031-34530-2_10

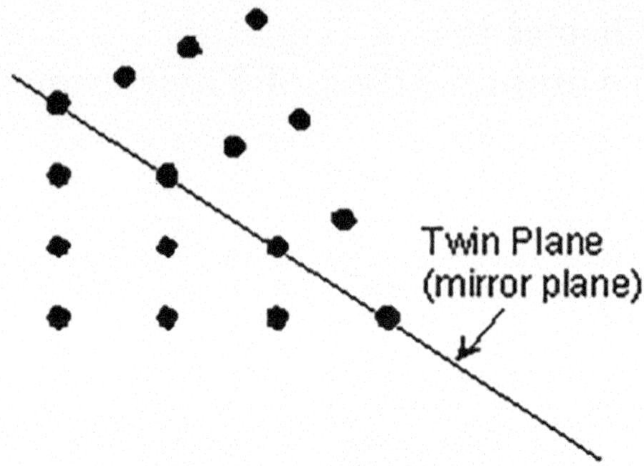

Fig. 10.1 Growth twin. Prof. S. A. Nelson, Twinning, Polymorphism, Polytypism, Pseudomorphism, Tulane University, 21, January 2019. EENS 2110, mineralogy. Open access

(1) Growth Twins-During crystal growth when some accidents occur and a new crystal is added to the face of an already existing crystal, twinning can occur if the new crystals share lattice points on the face of the existing crystal but has an orientation different from the original crystal. See a schematic illustration is seen in Fig. 10.1.
(2) Transformation Twins-Transformation twinning occurs when a preexisting crystal undergoes a transformation due to a change in pressure or temperature. Thus, when such changes occur, so that new crystal structures or parts of it become arranged in different symmetrical orientations, twins form during cooling (namely, when a decrease in temperature occurs).
(3) Deformation twins-In deformation twins formation atoms are displaced from their original location to produce a new symmetrical arrangement-the deformation twins-. Example 10.2 illustrates the process, in a) the original arrangement of the atoms, and in b) the twin configuration resulting from the deformation:

This introduction can be summarized with a commonly known note, e.g., mistakes are common during crystal growth in particular in stacking new planes relative to previously formed planes. Such defects in stacking sequence can produce within the crystal distinct regions called domains. Boundary separates two domains of a crystal. If the two domains of a crystal are mirror images (see Fig. 10.2b) they are known as twins, and the boundary that separated the domains are called twin planes.

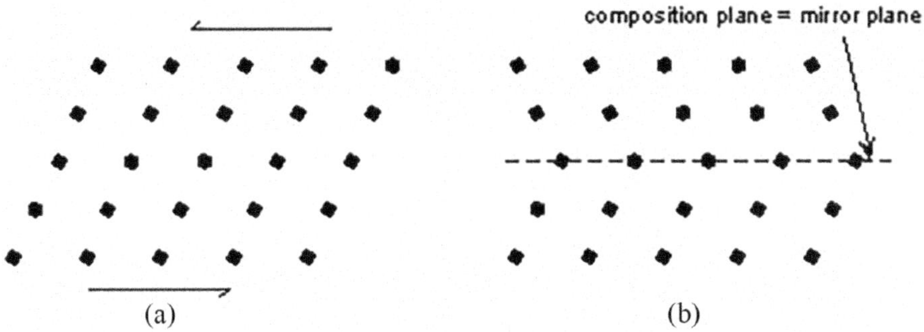

Fig. 10.2 Deformation twin. Prof. S. A. Nelson, Twinning, Polymorphism, Polytypism, Pseudomorphism, Tulane University, 21, January 2019. EENS 2110, Mineralogy. Open access

10.2 The Effect of Oxygen

10.2.1 In Si

Oxygen ion implantation of silicon is used in silicon base devices for the electronic modification of silicon-based devices. The implantation results in a material of reduced lifetime due to defects created by particle irradiation. Subsequent annealing can minimize the damage. Devices fabricated on thin silicon layers separated from their bulk substrates by a thin insulating layer, significant improvement of their performance is obtained. TEM illustrations are presented to show the defect structure developed during implantation, which are strongly affected by implantation temperature.

A set of (100) Si wafers was implanted with a high-current (1 mA / c m^2) oxygen implanter at 200 keV to a dose of $1.8 \times 10\ \text{cm}^{-2}$, at temperatures of 450, 500, 525, 550, 575, 600, and 700 °C. The final defect densities of the samples were determined after annealing in an argon atmosphere at 1300 °C for 6 h. Cross-section samples were examined by TEM with a Philips 400 T at 120 keV and also by HREM along the (110) axis in a JEOL 200CX and a JEOL 2000FX at 200 keV. The wafer temperature is an important parameter on defect density and type. Micro-twins and stacking faults were created in the top silicon layer during implantation from 450–550 °C. From 600–700 °C, numerous multiply faulted defects were observed. They are 40 to 140 nm long and consist of several discontinuous stacking faults which are randomly spaced. In the substrate region beneath the buried oxide layer created by implantation, the defects observed included stacking faults and {113} defects. Some parts of the {113} defects show a twinning structure across {115} planes. High dislocation densities were seen in samples implanted at 450–575 °C after annealing at 1300 °C for 6 h.

Electronic modification of silicon by ion implantation has been recorded. Due to defects created by the implantation the lifetime in a material is reduced, however subsequent annealing can minimize the damage. Significant improvement can be achieved if devices can be fabricated on thin silicon layers separated from their bulk substrates by a thin insulating layer. Compared to bulk silicon, integrated circuits fabricated on silicon-on-insulator material offer the advantages of higher speed, higher circuit density, higher temperature tolerance, and radiation hardness. Siliconon-insulator structures can be produced by implantation of a high dose of oxygenSIMOX (separation by implantation of oxygen). Recall that SIMOX stands for Single- Molecule Oxygen Complex and is made from pure silicon on insulator.

Figure 10.3 shows TEM micrographs of a SIMOX sample. It generally consists of three regions: a top Si layer about 0.μm thick; a buried oxide layer about 0.4 μm thick; and a mottled, defect-rich region below the oxide about 0.5 μm thick. Oxygen filled bubbles are observed which decrease in size as they extend downward from the surface. Bubble formation is temperature dependent and are generated at about 550 °C. In the lower portion of the top silicon layer, there are small stacking faults and {111} microtwins. At the bottom layer in Fig. 10.3b linear defects of 40–140 nm long lying along {111} planes can be seen and referred to as MFD (multiple faulted. Defects) which consists of discontinuous stacking faults. It consists of intrinsic (I) and extrinsic (E) stacking faults separated by 6 atomic layers, which were created from a shear mechanism, and the stresses can be relieved by shear formation during annealing which are composed of partial dislocations which, in the temperature range of 525–700 °C, are sufficiently mobile to move and shear the silicon crystal. The atomic layer of the linear defects was inspected by HREM and illustrated in Fig. 10.4. Random faulting of MFDs, rather than a regular twinning arrangement, may be due to the presence of several inhomogeneous nucleation-sites along the edge of the same defect. Beneath the buried oxide layers are damaged zones which contain numerous {113} defects and stacking faults. The defects may consist of hexagonal silicon due to self-interstitial precipitation and is illustrated in Fig. 10.5. Defects {113} of Fig. 10.5a resembles hexagonal silicon, while (b) shows a high-resolution image of another {113} defect.

The interfaces between the defect and the matrix, are highly distorted because they must accommodate the strain of the defect. When the samples are annealed at 1300 °C for 6 h, the near-surface bubbles and the MFDs are eliminated. Annealing samples implanted at 525, 550, and 575 °C results in a final density of stacking faults and threading dislocations of 10^9 cm^{-2}. Figures 10.6a and 4 (b) show a cross-sectional view and a plan view, respectively, of a sample implanted at 550 °C after annealing at 1300 °C for 6 h.

In summary, TEM indicate that in implantation and annealing the temperature has the strongest effect on the microstructure and the defect formation in SIMOX materials. In samples implanted from 450–550 °C stacking faults and microtwins form. The {113} defects are observed in the substrate region at all temperatures and are composed of different structures, including hexagonal and twinned cubic diamond silicon. Samples implanted at higher temperatures show after annealing lower defect densities than those implanted at lower temperatures.

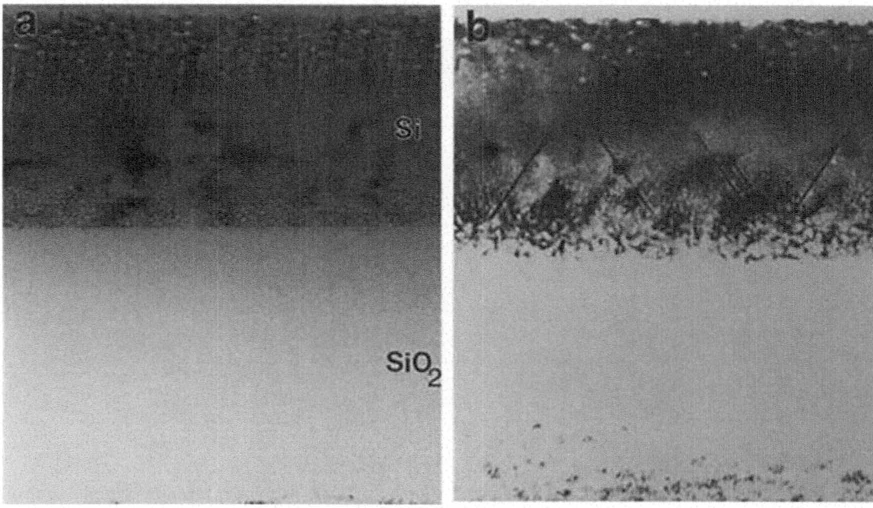

Fig. 10.3 A cross-sectional view of SIMOX structure which consists of three regions: a top Si layer; a buried oxide layer; and a mottled, defect-rich region below the oxide. **a** sample implanted at 550 °C and b sample implanted at 600 °C. Note the linear defects in the top silicon layer in **b**. S. Seraphin, Solar Energy Materials and Solar Cells 32, 343 (1994). With kind permission of Elsevier

10.2.2 In SiC

Most of the data reported on twinning in SiC are related to nano structures, which comprises a section in the chapter on nanomaterials. However, here a nanostructure *is* considered with length up to several hundreds of microns and with relatively homogeneous diameters. The synthesis of twinned SiC nanowires was carried out in a high-temperature vacuum furnace by a catalyst-free thermal CVD method. An ntype Si (100) wafer was used as the substrate.

The phase composition, morphology, and the microstructure of the as received SiC nanowires were characterized by XRD, field emission microscopy (FESEM), EDS, TEM and HRTEM. The as-synthesized twinned nanowires were up to several hundred microns long and had relatively homogeneous diameters in the range of 20,100 nm. The small amount of oxygen comes from the surviving thin oxide shell and remained as oxygen in the system, and thus twinning occurred in oxygen environment. The general morphology of the CVD SiC nanowires grown on Si substrate was investigated by FESEM and the images are shown in Figs. 10.7a at low magnification and in Fig. 10.7b at high magnification. As seen both straight and curved nanowires grew randomly on the Si substrate. Additional details can be seen in Fig. 10.8a, b, c, d, e, f by TEM and HEREN of the structure and morphology of the as-fabricated SiC nanowires.

In summary, small amount of oxygen comes from the fabrication process. The twinned SiC nanowires were synthesized by a CVD method on silicon wafer substrate. The as-obtained twinned nanowires had lengths of up to several hundred

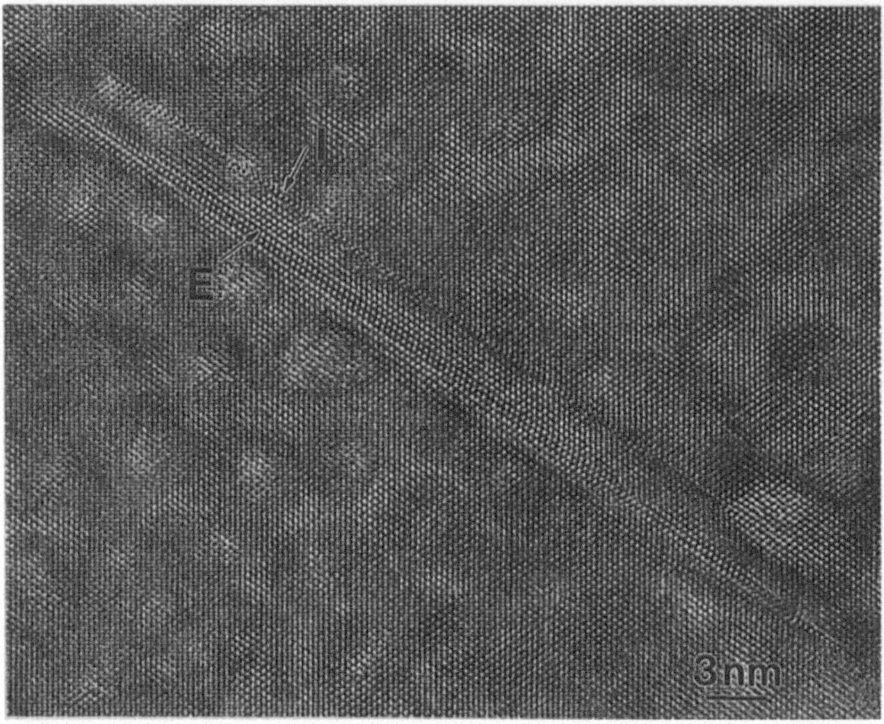

Fig. 10.4 A high-resolution electron micrograph of a multiply faulted defect. It is 80 n m long and 12 atomic layers wide. It consists of intrinsic (I) and extrinsic (E) stacking faults without an extra plane inserted along the defect plane. This suggests that the defect was created from a shear mechanism rather than from the accumulation of the excess interstitials. Seraphin, Solar Energy Materials and Solar Cells 32, 343 (1994). With kind permission of Elsevier

microns and relatively homogeneous diameters in the range of 20–100 nm. The vapor solid (VS) growth mechanism was responsible for the initial nucleation and.

Huang, S. Liu, S. Huang, Yan-gai Liu and X. Wu, RSC Adv., 4, 18,360 (2014) twinned nanowire formation. A competition of stacking fault energy and electrostatic energy was suggested to explain the growth of stacking faults and twinning. The twin structure was energetically favorable to the growth of SiC nanowires.

10.2.3 In GaAs

Growth of GaAs (111) epilayer on Ge (111) wafer was performed by MBE epitaxy (The epitaxy system operates in ultra-high vacuum). The (111)A orientation GaAs was homogeneous as measured by TEM and high energy electron diffraction.

Twinned domains are the main defects in our GaAs (111) epilayers.

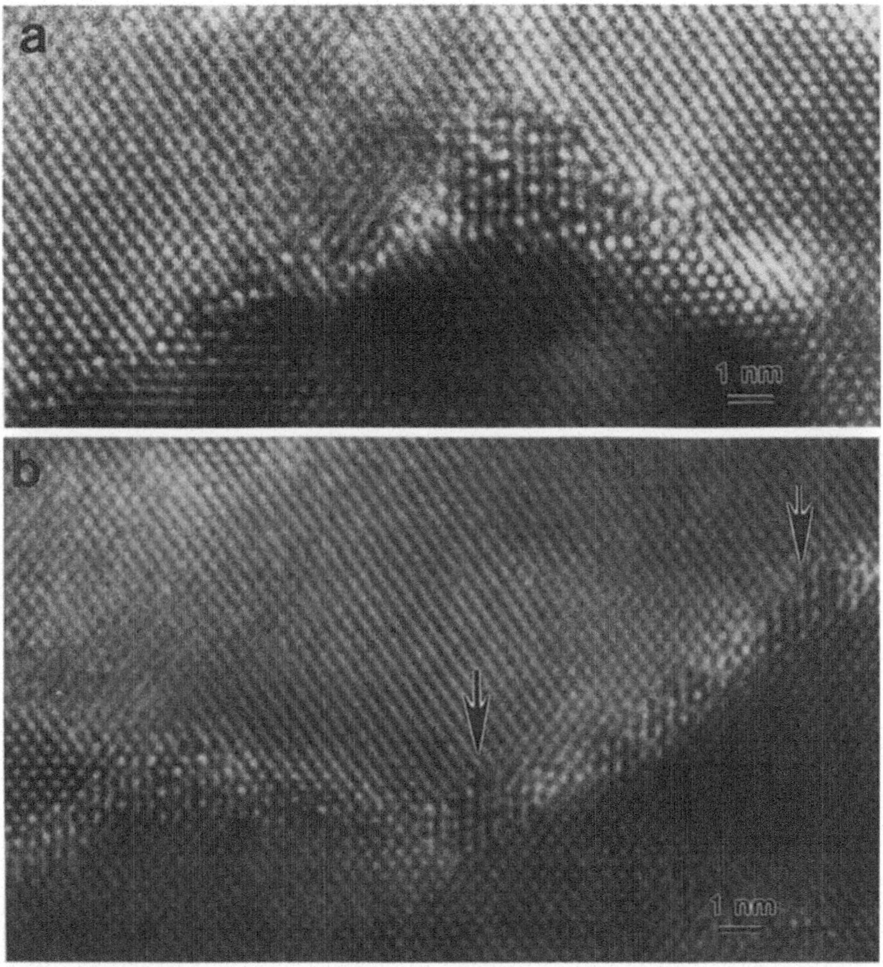

Fig. 10.5 **a** A high-resolution TEM image of a {113}defect showing an atomic structure of hexagonal silicon. **b** A high-resolution image of another {113} defect. It illustrates a zigzag character of the defect. Arrows indicate regions of defect exhibiting crystalline structure of twins across the {115}planes. Seraphin, Solar Energy Materials and Solar Cells 32, 343 (1994). With kind permission of Elsevier

The chemical analysis of the epilayer using TEM–EDX, S1, does not reveal any increased oxygen and the oxygen content is homogeneous over the whole cross section. Reflection High Energy Electron Diffraction (RHEED) was sed to characterize the surface reconstructions. Figure 10.9 is an illustration of the system exhibiting the Ge substrate and the GaAs before and after growth. Different ratios of beam equivalent pressure (BEP) of BEP_{As4}: BEP_{Ga} were used (31:1 to 94:1) to get GaAs epilayers, which is illustrated in Fig. 10.10 by optical microscopy and SEM. Optical micrographs and TEM of the respective samples are shown in Fig. 10.11.

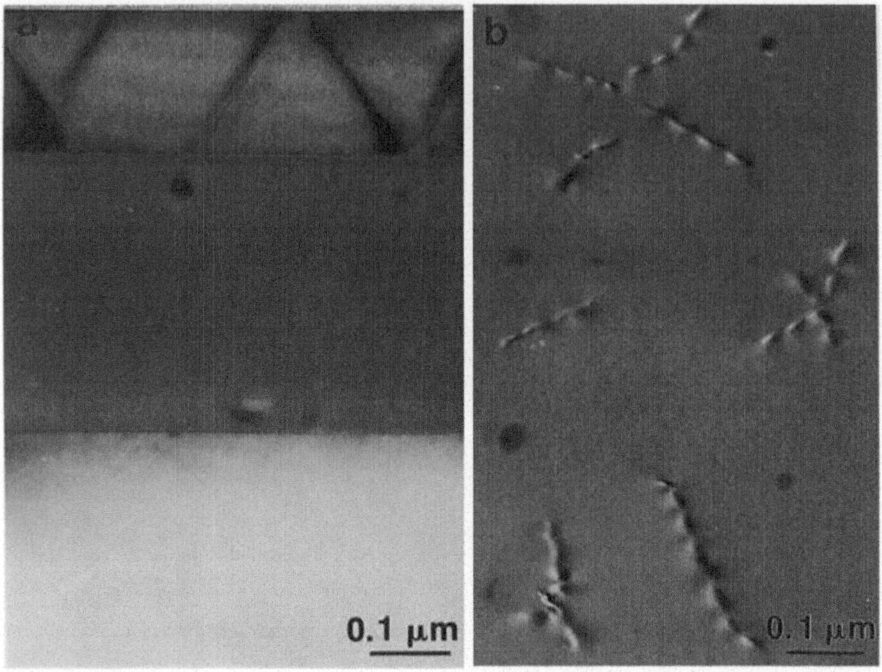

Fig. 10.6 **a** A cross-sectional view of sample implanted at 550 °C after annealing at 1300 °C in argon atmosphere for 6 h. **b** A plan view of the top silicon layer showing a high defect density of 10^9 cm^{-2}. Seraphin, Solar Energy Materials and Solar Cells 32, 343 (1994). With kind permission of Elsevier

Orientation of GaAs presented in Fig. 10.12, confirming twinning by the use of SEMEBSD.

The effect of twin boundaries is detrimental on the layer's optical quality, and it is very important to limit the density of twinned domain in GaAs (111) epilayers. Twinning is a common problem in the heteroepitaxy of (111) oriented III-V compounds. The assessment of the epilayer quality with respect to twin defects is not easy. XRD only indicates the volume fraction of the twinned phase, but not the density which is the critical parameter.

There is an optimum for the BEP_{As4}: BEP_{Ga} ratio for twinning. At low As:Ga ratio (31:1) the surface is rough, while it is much smoother and slightly less twinned at intermediate As:Ga ratio (62:1). It has been shown that small twinned domains form at 600 °C and low As:Ga ratio, while no twins are observed for higher As:Ga ratios. After the initial improvement with increasing V:III ratio (31:17 → 62:1), a degradation of the surface morphology occurs. A likely reason of the degradation of the GaAs/Ge (111) interface is at the initial stages of the growth, which leads to the growth of large twinned domains.

In summary, the growth of GaAs (111) epilayers on Ge(111) wafers by MBE was considered in this section. The oxygen content did not increase during the fabrication

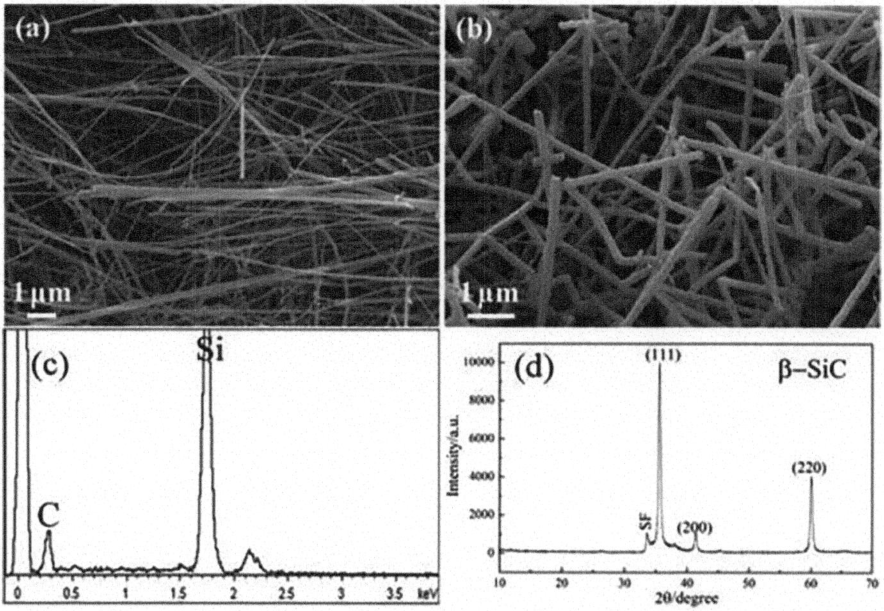

Fig. 10.7 **a** Low magnification and **b** high magnification FESEM images of the asgrown nanowires; **c** EDS pattern of the nanowires; **d** X-ray diffraction pattern of the nanowires. Z. Huang, H. Liu, K. Chen, M. Fang, J. Huang, S. Liu, S. Huang, Yan-gai Liu and X. Wu, RSC Adv., 4, 18,360 (2014)

and it is homogeneous over the whole cross section of GaAs. The polarity of GaAs is (111)A and remains homogeneous over the full area, as measured by TEM and high energy electron diffraction. The main defects observed in GaAs (111) epilayers are twin domains. Only a narrow domain of As:Ga ratios lead to the growth of smooth and twin-free GaAs (111)A epilayers during the growth. At low As:Ga ratio, the surface is rough; while at high As:Ga ratio the epilayers present large densities of twinned domains.

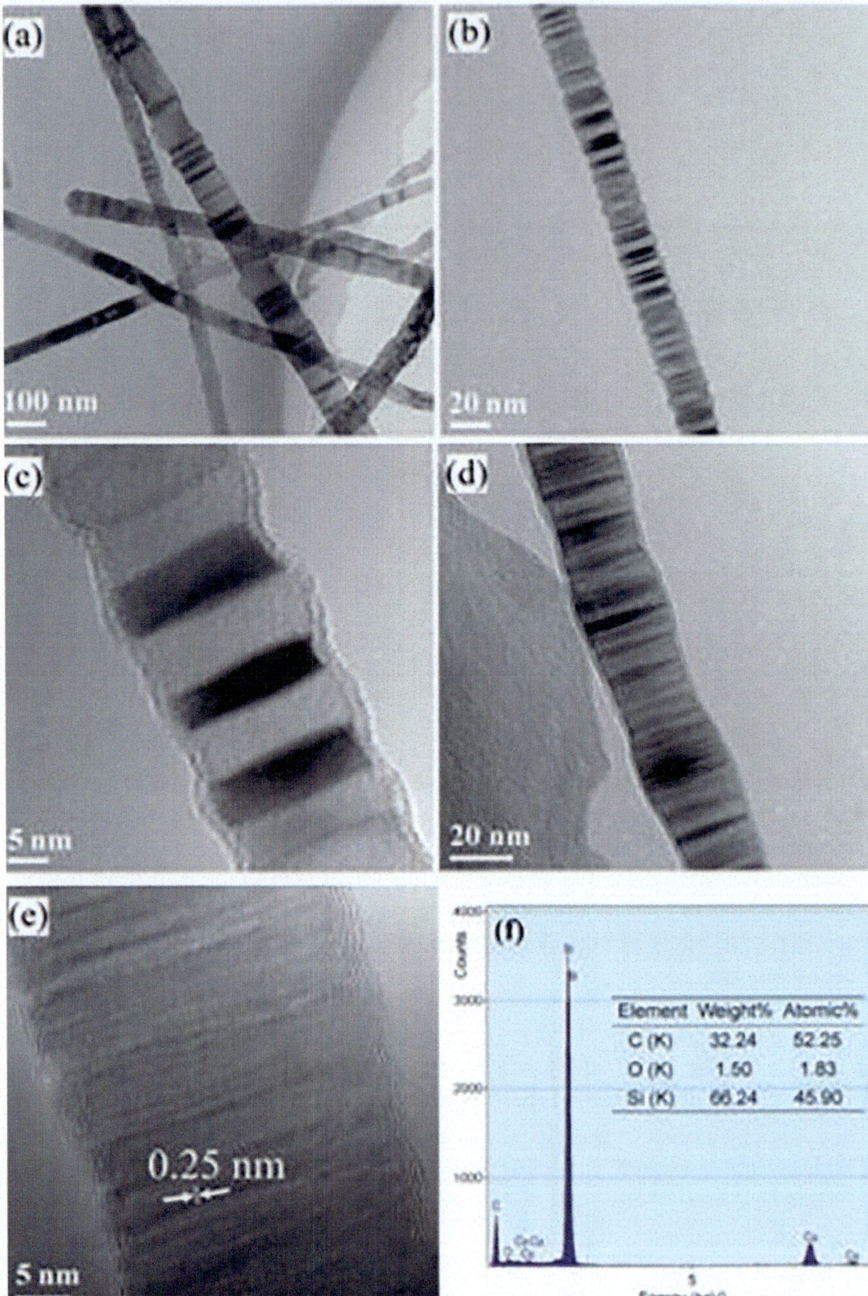

Fig. 10.8 **a**, **b**, **c**, **d** Typical low magnification TEM images of the twinned SiC nanowires; (**e**) a typical HRTEM image of the twinned SiC nanowires; **f** the corresponding EDS of the nanowires. Z. Huang, H. Liu, K. Chen, M. Fang, J

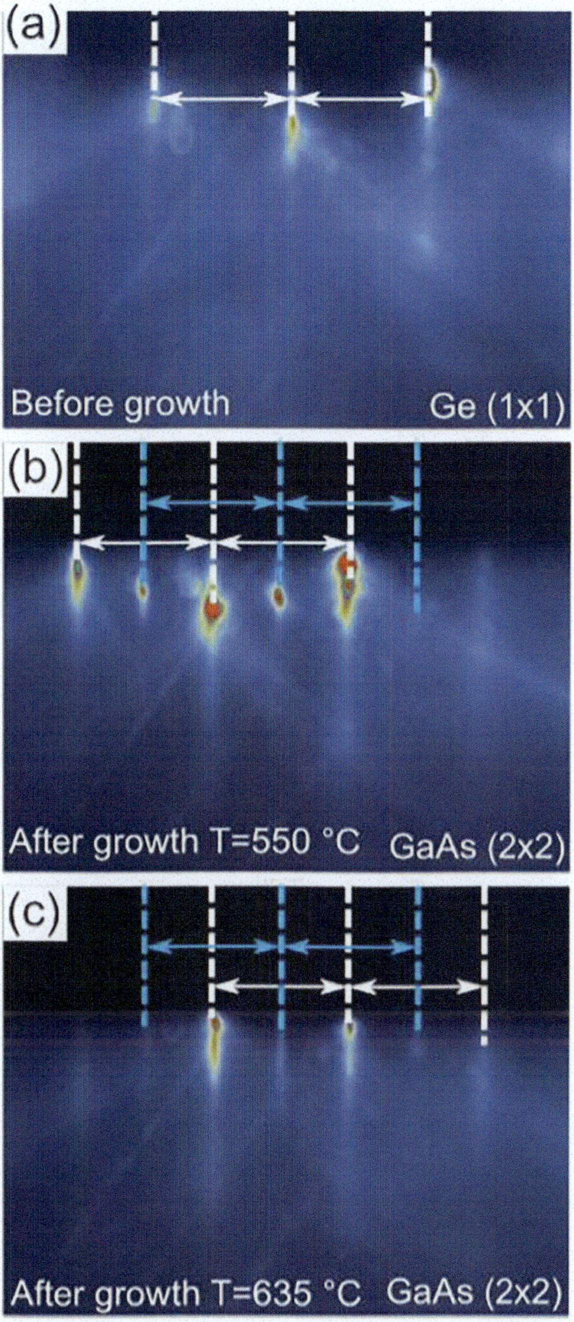

Fig. 10.9 RHEED characterization before and after GaAs growth (~9 nm thick) using 62:1 BEP ratio. D. Pelati, G. Patriarche, O. Mauguin, L. Largeau, L. Travers, F. Brisset, F. Glas, F. Oehler, J. Crystal Growth, Elsevier, 519, 84 (2019). Open Science

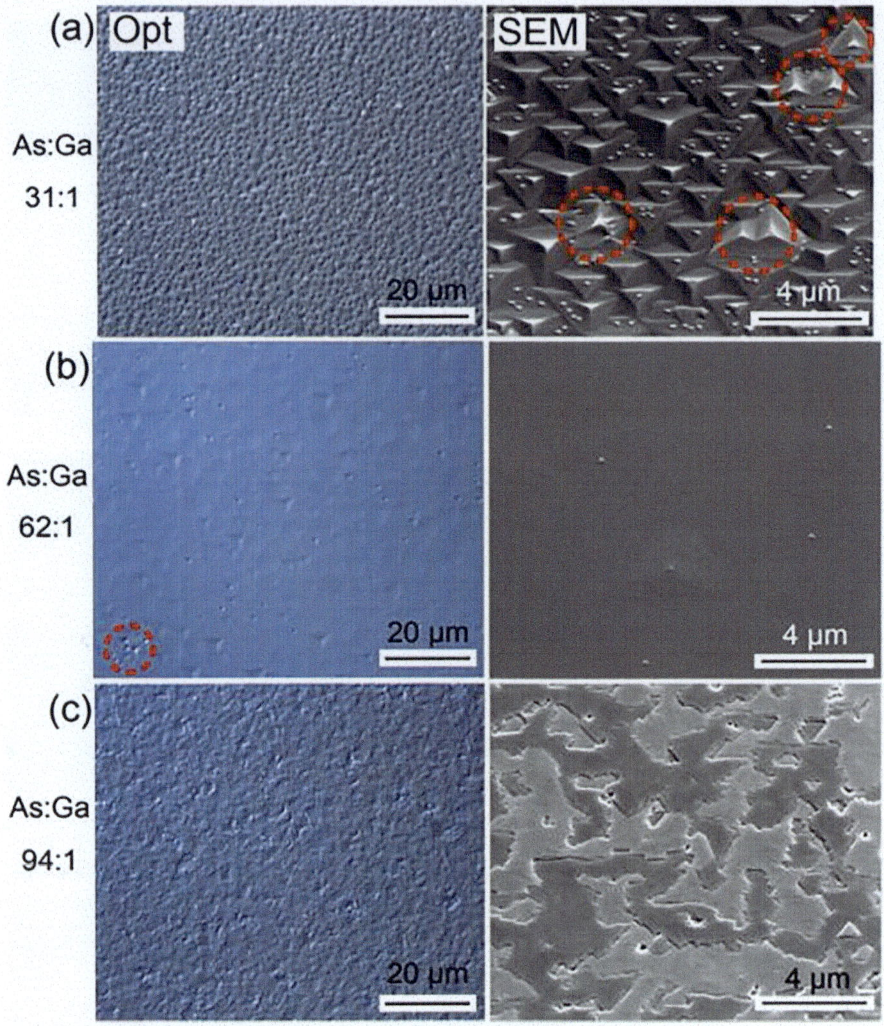

Fig. 10.10 Optical microscopy (Nomarski, left) and SEM view (45° tilt, right) of the surface of GaAs epilayers (~450 nm) on Ge (111) grown with different As:Ga BEP ratios. (a) 31:1 (b) 62:1 (c), 94:1. D. Pelati, G. Patriarche, O. Mauguin, L. Largeau, L. Travers, F. Brisset, F. Glas, F. Oehler, J. Crystal Growth, Elsevier, 519, 84 (2019). Open Science

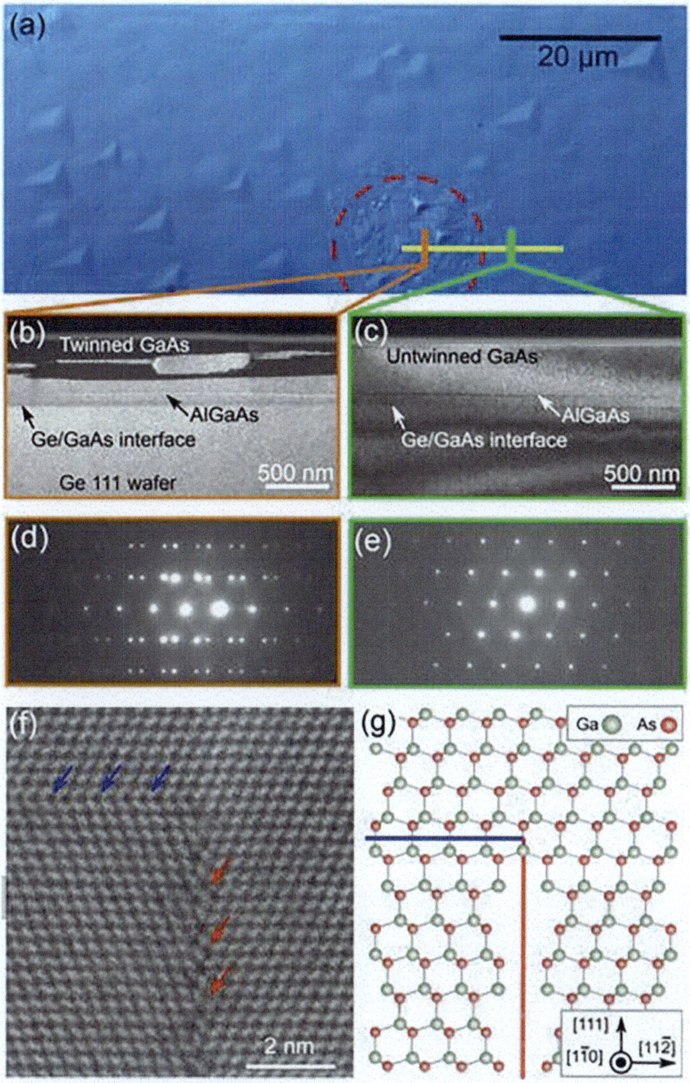

Fig. 10.11 **a** Optical micrograph of the zone selected for cross-section TEM. The yellow line indicates the position of the FIB cut, across a rough patch with 180° rotated pyramids (red circle). **b** Dark-field TEM from the rough area, orange rectangle in (a), showing twin contrast. **c** Dark-field TEM from the smooth region, green rectangle in (a) observed in the same conditions. **d** TEM diffraction pattern near the [110] zone axis in the twinned area. **e** TEM diffraction pattern in the smooth area. **f** Atomically resolved STEM-HAADF image of a twin boundary. The (111) plane is indicated by blue arrows and the vertical defective boundary by red arrows. **g** Corresponding atomic structure for two twinned GaAs cubic crystals. D. Pelati, G. Patriarche, O. Mauguin, L. Largeau, L. Travers, F. Brisset, F. Glas, F. Oehler, J. Crystal Growth, Elsevier, 519, 84 (2019). Open Science

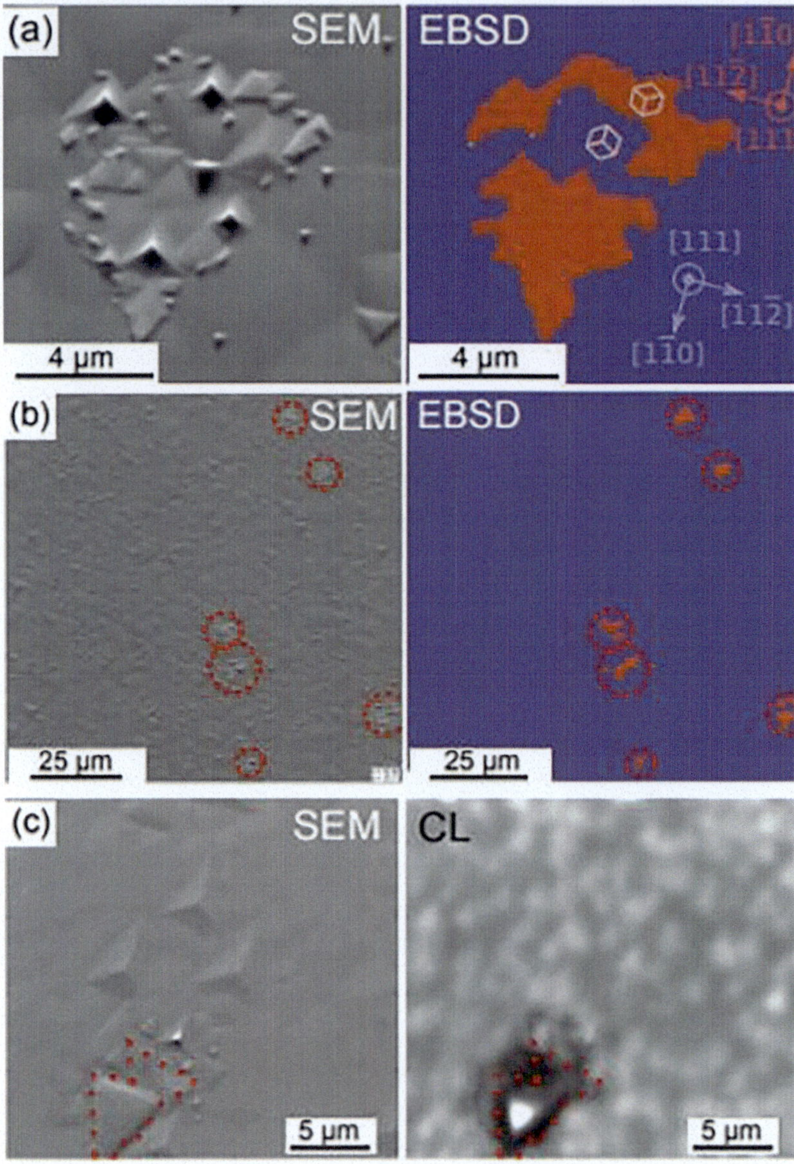

Fig. 10.12 **a** SEM-EBSD characterization of a small rough area showing the emergence of two small twinned domains (orange) in the main layer (purple). The orientation of the crystal is indicated by a color relative to its Euler angles with the EBSD detector. The projected cubic unit cell is shown as a pink polygon. Crystal orientations are labeled with matching colors. **b** Large scale SEM-ESBD characterization. Rough and twin area are indicated by red circles. The EBSD color scale is the same as in (a). (c) Room temperature SEM-CL (6 kV) of a small twinned area. The CL color scale is proportional to the integrated GaAs peak intensity. The dotted red line indicates the estimated position of the defective twin boundary. D. Pelati, G. Patriarche, O. Mauguin, L. Largeau, L. Travers, F. Brisset, F. Glas, F. Oehler, J. Crystal Growth, Elsevier, 519, 84 (2019). Open Science

References

Z. Huang, H. Liu, K. Chen, M. Fang, J. Huang, S. Liu, S. Huang, Yan-gai Liu and X. Wu, RSC Adv., **4**, 18360 (2014)

D. Pelati, G. Patriarche, O. Mauguin, L. Largeau, L. Travers, F. Brisset, F. Glas, F. Oehler, J. Crystal Growth, Elsevier, 519, 84 (2019). HAL Id: hal-02351877

Prof. S. A. Nelson, Twinning, Polymorphism, Polytypism, Pseudomorphism, Tulane University, 21, January (2019). EENS 2110, Mineralogy

S. Seraphin, Sol. Energy Mater. Sol. Cells **32**, 343 (1994)

Chapter 11
Fracture Toughness–Effect of Environment

11.1 Introduction

Fracture toughness expresses the resistance of materials -generally brittle- to the propagation of flaws under an applied stress. Contrary to brittle materials, plastic materials undergo observable plastic deformation and absorb considerable amount of energy before fracture. Deformation in brittle materials is characterized by low or almost nil energy prior to breaking. The consequence usually, is crack formation which propagates very fast mostly without the increase of stress. The brittle crack is perpendicular to the stress direction. It is customary to distinguish between three stages or three modes of fracture toughness as usually are known: Mode I-Opening mode (a tensile stress normal to the plane of the crack), Mode II-Sliding mode (a shear stress acting parallel to the plane of the crack and perpendicular to the crack front), and. Mode III-Tearing mode (a shear stress acting parallel to the plane of the crack and parallel to the crack front). Schematically these modes are shown in Fig. 11.1

Ceramics and semiconductors are characterized by brittle fracture, and brittle cracking starts generally at preexisting defects in the material. On applying a load, the preexisting defects are locations of stress concentration which can exceed the cohesive strength of the material. When this occurs cracks form and propagate until failure sets in. The stress field is related to the shape, size of defect, i.e., crack and the applied load. The stress intensity factor concept, K, is commonly accepted for the analysis of fracture toughness and it is associated with the load and the dimensions and other geometrical features of the crack (defect). More specifically the stress intensity factor is denoted by K_I. The maximum value for cracking is denoted K_{Ic}, which is defined as the plain strain fracture toughness. It is a measure of a material resistance to crack extension under predominantly linear-elastic conditions. Under low toughness condition little or no plastic deformation is occurring at the crack tip. A test specimen must be of large enough dimensions to give predominantly plane strain conditions at the crack tip. This depends on the specimen's initial crack length,

J. Pelleg, *The Influence of Environment on Mechanical Characteristics of Key Semiconductor Components*, https://doi.org/10.1007/978-3-031-34530-2_11

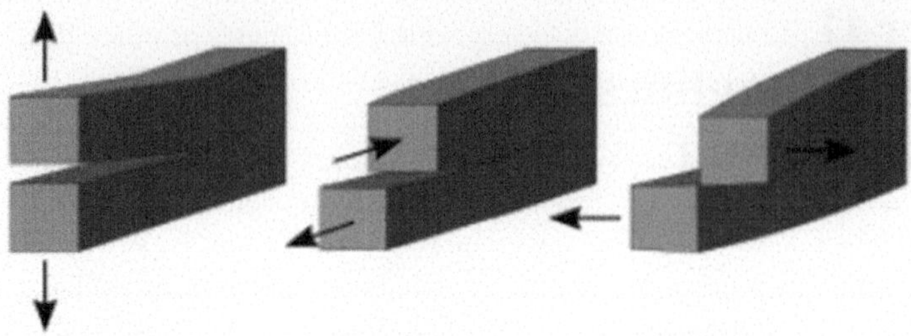

Fig. 11.1 Three stages or three modes of impact test. From WikipediA The Free Encyclopedia (Boniecki et al., 2016)

thickness, and size, and also on the yield strength of the material being tested, a crack for example. The accepted expressions of Griffith Boniecki et al., 2016 on critical crack which must satisfy the criterion for the crack lengths, the so-called Griffith criterion, is expressed as

$$\sigma^2 \pi a = 2\gamma_s E \tag{11.1}$$

where "a" is the length of the crack, E is the Young's modulus, γ_s is the specific (per unit area) energy associated with the new interfaces being created and σ is the applied stress. Further,

$$\frac{\sigma^2 \pi a}{E} = G_C \tag{11.2}$$

where G_C is the resistance to cracking the toughness. For the elastic-brittle case from (11.1) and (11.2)

$$G_C = 2\gamma_S \tag{11.3}$$

Irwin placed the focus on the stress–strain distribution in the vicinity of the crack tip. K_I provides a quantitative measure used to assess the propensity of a crack to grow and expressed in terms of a stress intensity factor. A critical value of this parameter, K_{IC}, can be assumed a material property, and is called fracture toughness. For plane stress state

$$\sigma^2 \pi a = G_C E = K_{IC}^2 \tag{11.4}$$

According to the Linear Elastic Fracture Mechanics crack extension occurs when the stress intensity factor K_I expressed in terms of the stress σ and crack (half-)length *a*, reaches the critical value of the stress σ and crack (half-) length a, reaches the critical value K_{Ic}, thus K_I:

$$K_I = Y\sigma\sqrt{\pi a} = K_{IC} \tag{11.5}$$

the stress intensity factor can be expressed in units of MPa 1m$^{1/2}$ and Y is a function related to the crack length.

It may be noted that for fracture toughness, evaluated by indentation method of radial cracks at the edge of Vickers indentation K_c is

$$K_C = \alpha\sqrt{\frac{E}{H}\frac{P_{max}}{C^{3/2}}} \tag{11.6}$$

where c is the average crack length, H the hardness of the material and P_{max} is the maximum load during indentation. For brittle materials α is about 0.061. There is considerable interest in determining the fracture toughness of brittle materials by measuring the extent of cracking associated with a Vickers indentation because of the ease of specimen preparation and the simplicity of the test.

All the usual test methods, such as tension, torsion bend or indentations can be principally used to evaluate fracture toughness, but the most common toughness test is the impact test in particularly the Charpy impact test, due to the ease of specimen preparation and the simplicity of the test. Because the frequent choice to use impact test to evaluate fracture toughness, and since the description of the other tests (although not for toughness measurements) earlier in the book, a presentation of the impact test is added to get acquainted with the procedure.

11.2 Impact Test

There are two main forms of impact test, **the Izod and the Charpy test**. Both involve striking a standard specimen with a controlled weight pendulum travelling at a set speed. The amount of energy absorbed in fracturing the test piece is measured and this gives an indication of the notch toughness of the test material. Both Charpy and Izod impact testing are popular methods of determining impact strength, or toughness, of a material. In other words, these tests measure the total amount of energy that a material is able to absorb. This energy absorption is directly related to the brittleness of the material. The Izod impact testing holds the sample vertically with the notch facing the pendulum. The Charpy impact testing holds the sample horizontally with the notch facing away from the pendulum. It indicates the amount of energy needed to break standard test specimens under specific conditions of specimen, mounting, notching and pendulum velocity at impact. The Izod test is most commonly used to evaluate the relative toughness or impact toughness of materials and as such is often used in quality control applications where it is a fast and economical test. The notch in the sample affects the results of the impact test, thus it is necessary for the notch to be of regular dimensions and geometry. The size of the sample can also affect results, since the dimensions determine whether or not the material is in plane strain. While the Charpy impact test, also known as the Charpy V-notch test, is a high strain-rate test that involves striking a standard notched specimen with a standard weight pendulum swung from a set height. The impact test helps measure the amount of energy absorbed by the specimen during fracture.

11.2.1 Notch Effect

A general term of notch is basically a change in the cross section of a component arising from grooves, drill holes or any variation in the cross section of a material. To investigate its effect on the performance of a component a notch is artificially introduced into a component. Regarding the common use of notch in engineering it is usually a cut (slit) in the sample and its form depends on the shape of the specimen (if round the notch is around the periphery). Notches result in uneven stress distribution, leading to stress peaks -known as the notch effect. Because the notch effect, notch is a very important geometry with widespread applications in engineering structural components. In particular -due to its being a source of stress concentration- it plays an important role during safety designing of engineering components. Thus, it is of great significance understanding and assessing the sensitivity of components to geometrical discontinuities naturally present or intentionally introduced into the samples under investigation. In engineering and materials science, a notch refers to a deliberately introduced V-shaped, u-shaped, or circular defect in a planar material. Further, it is important to be acquainted also with the concept of the ratio of the fatigue strength of an unnotched specimen to the fatigue strength of a notched specimen of the same material and condition; both strengths are determined at the same number of stress cycles.

As indicated above in structural components, a notch causes a stress concentration which can result in the initiation and growth of crack in particular fatigue cracks.

Notches are used in materials characterization to determine fracture mechanics related properties such as fracture toughness and rates of fatigue crack growth. Notches are commonly used in material impact tests where a morphological crack of a controlled origin is necessary to achieve standardized characterization of fracture resistance of the material. The most common is the Charpy impact test, which uses a pendulum hammer (striker) to strike a horizontal notched specimen. Impact energy is a measure of the work done to fracture a test specimen. The test specimen continues to absorb energy and work hardens at the plastic zone at the notch. When the specimen can absorb no more energy, fracture occurs. The Izod impact strength test uses a circular notched vertical specimen in a cantilever configuration.

Impact test is the ability of a material to withstand impact, used by engineers to predict its behavior under actual conditions. Many materials fail suddenly under impact, at flaws, cracks, or notches.

Three basic factors contribute to a brittle-cleavage type of fracture. They are:

1. a triaxial state of stress
2. a low temperature, and

3. a high strain rate or rapid rate of loading.

The three factors do not have to act or be present at the same time to induce brittle fracture. Generally, a triaxial state of stress exists at the notch, and low temperature are responsible for most of brittle type failure in service. However, since these effects are accentuated at a high rate of loading, many responsible for most service failures of the impact tests have been used to determine the susceptibility of materials to brittle behavior. A schematic illustration of an impact machine and the shape of a notched specimen are shown in Fig. 11.2 reused from TWI Global (Global and company n d). The standard Charpy-V specimen, illustrated in Fig. 11.3 is 55 mm long, 10 mm square and has a 2 mm deep notch with a tip radius of 0.25 mm machined on one face.

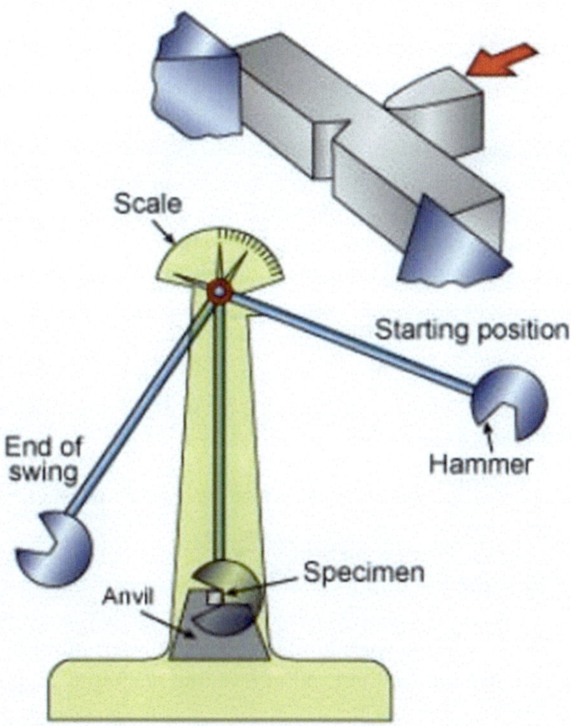

Fig. 11.2 Charpy testing machine and test specimen. From TWI global

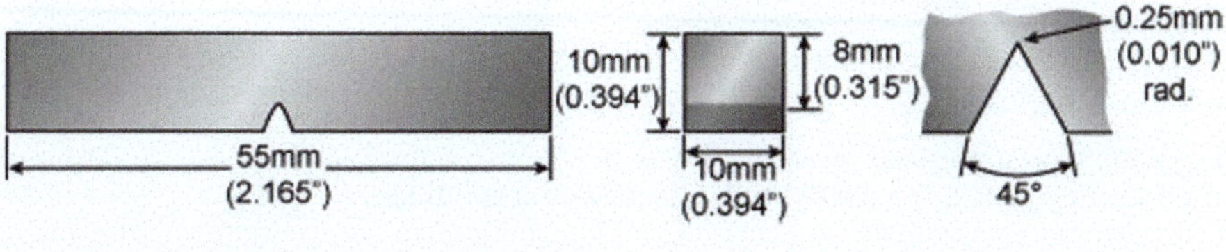

Fig. 11.3 Standard Charpy-V specimen. TWI global

11.3 The Effect of Oxygen

11.3.1 In Si

Cracking in brittle materials, such as ceramics and single crystal semiconductors, sets in without or with minimal plastic deformation usually from preexisting defects in the material (either on the surface or in the bulk). The stress concentration at defects (pores, microcracks, inclusions, etc.) exceeds the applied force in defect free material which is usually the case. If the stress concentration exceeds the cohesive strength of a material, crack forms which induces failure in the material. Thus, the state of the stress field near the edges of the existing defect results in fracture according one of the modes illustrated earlier in Fig. 11.1. The stress field is related to the shape and the size of the defect, the crack, and to the type and the magnitude of the applied load. The mechanical properties of the silicon single crystal, in particular the fracture toughness is considered below. The silicon was obtained by the floating zone (FZ) technique and has a crystal orientation < 100 > and a resistivity of ~ 2000 Ωcm. Bending strength was used to evaluate the fracture toughness at 20 °C and 700 °C. In addition, the Young's modulus and material hardness have been determined. The concentration of interstitial oxygen atoms was equal to ~ 5×10^{15} cm^{-3}. Note, that based on literature data, it has been indicated that clusters of oxygen atoms or SiO_x precipitates (inclusions) are those defects in the structure of Si, from which the critical crack can start to propagate. Further note that the values of K_{Ic} and the bending strength at 20 and 700 °C are respectively, $K_{Ic} = 1.3 \pm 0.1$ $MPam^{1/2}$ and the four-point bending strength $\sigma_c = 289 \pm 61$ MPa, and at 700 °C $K_{Ic} = 20 \pm 3$ MPa $m^{1/2}$ and $\sigma_c = 592 \pm 86$ MPa. Three- and four-point bending strength, σ_c is expressed as

$$\sigma_C = \frac{1.5P_C(L - l)}{bw^2} \tag{11.7}$$

P_c is the fracture load. L is the spacing of the lower support, l is the distance of upper pressure rolls, b is the sample width and w is the sample thickness.

For $l = 0$, the sample undergoes the three-point bending. Equation (1.5) written differently for the three-point bending test[5] is

$$K_{Ic} = Y\frac{1.5p_C L}{bw^2}c_k^{0.5} \tag{11.8}$$

where Y is a geometrical constant and c_k is the notch depth. When using indentation by Vickers hardness, K_{Ic} is given (Irwin, 1957) by

$$KIc = 0.016\,(HE)0.5\,(cP1.5) \tag{11.9}$$

where E is Young's modulus, H is the Vickers hardness, P the load of the indenter and c is propagating from the corners of the length the imprint.

References

M. Boniecki, P. Kamiński, W. Wesołowski, K. Krzyżak, Materialy Elektroniczne (Electronic Materials) **44**, 42 (2016)

WikipediA The Free Encyclopedia, wikipedia.org/wiki/Fracture mechanics

TWI Global company, uk

G.R. Irwin, J. Appl. Mech. **24**, 361 (1957)

Chapter 12
Deformation in Nanostructure–Effect of Environment

12.1 Introduction

Dual aspects are considered in this chapter. On the one hand mechanical properties similar to those discussed in macro size specimen, on the other hand the environmental effects are discussed in the earlier chapters, focusing mainly on the effects of oxygen and moisture. Their effect on the overall properties is dependent on the type of the semiconductor, their intended application, and the conditions of their use. Thus, the behavior of oxygen in silicon is quite complex Pelleg (2021). Further, already the mode of the growth of silicon results in different amounts of oxygen and their distribution in the crystal. For example Pelleg (2021) Czochralski (CZ) grown silicon has different oxygen amount and distribution compared to the Float Zone (FZ) one. One has to expect also regarding the moisture, that their influence is a function of various factors.

The mechanical properties of nanomaterial have been presented recently in a book by Pelleg (2021) and interested readers on the subject are referred to consult it. In general nanomaterials exhibit drastically different properties compared to their bulk counterpart having the same composition. Recall, that generally a common definition of nanomaterials is related to a substance where at least one dimension is less than 100 nm thus in a scale comparable to atoms and molecules since 1 nm = 10^{-9} m.

Understanding the mechanical properties of nanomaterials is essential for their proper use in a variety of industrial applications, like electronic devices, sensors, thin films, composite and not least for medical purposes. One is pondering whether the environmental affects -such as oxygen, or moisture-have the same or similar impact on the general behavior of nanomaterials as in the case of macroscopic dimensions material. Since nanomaterials in modern industry are exposed to various technological applications where environmental contributors are present or even playing active role, it is clearly important to have as much information as possible on the behavior of various nanomaterials in actual use. As indicated in the preface of this book of the nanomaterials chosen as the subject of consideration are Si, Ge, SiC

J. Pelleg, *The Influence of Environment on Mechanical Characteristics of Key Semiconductor Components*, https://doi.org/10.1007/978-3-031-34530-2_12

and GaAs limited only by the availability in the literature. The various features and properties might be generally available, but having both, some mechanical property of interest with an environmental factor acting simultaneously is often not treated in the published literature. Such limitations are reflected in earlier chapters, and the present is not an exception.

12.2 Nano-Wear Deformation

12.2.1 Effect of Oxygen

12.2.1.1 In Si

The wear deformation in mono-crystalline silicon under nano-sliding is considered in this section. All experiments were conducted on the 100 surfaces. Sliding was generated by a commercially available sliding system illustrated in Fig. 12.1. The sliding counterpart of the silicon surfaces was a polishing pad, 1 mm in thickness and 15 mm in diameter, made from felt fibres Zhang and zarudi (1999). The elastic modulus of the pad, E_{pad}, as determined by a uniaxial tensile test on an Instron, with an elongation speed of 0.5 mm/min under a wet condition. Tests were repeated six times. Poisson's ratio was calculated by measuring the transverse deformation of the pad during elongation, which was done by a video camera recording system attached to the Instron machine. The magnification used was 100 times. The duration of each sliding test was 15 min.

The normal sliding loads applied were 147 mN, 294 mN and 441 mN, respectively. With the aid of an optical microscope, the wear rates under these loads were determined by measuring the volume removed in sliding and found to be 69 mm^3rs, 92 mm^3rs and 213 mm^3rs, respectively. The maximum depths of the worn cavities were 2 mm, 2.3 mm, and 3.5 mm, correspondingly. Damage free specimens before sliding experiments with Rms surface roughness of 3 nm is illustrated in Fig. 12.2. (note: The Rms average measures the average height deviations of the mean line. Rms is calculated by measuring the height of a surface's microscopic peaks and valleys, squaring those measurements, determining the average of the squares, and finding the square root of that number). This insured that sub-surface deformation and micro-structural changes of worn specimens were the result of sliding. After sliding (a worn specimen surface topography) the typical surface is seen in Fig. 12.3. It is free from micro-fracture, but 4–9 mm wide scratches up to 20 mm deep are seen as seen in Fig. 12.4. Atomic resolution sub-surface images are illustrated in Fig. 12.5 indicating that an amorphous silicon is always formed during sliding. The layer thickness clearly is expected to depend on applied sliding load. Under a load of 441 MN, the thickness of the amorphous layer was 2.7 nm, which decreased with decrease of the load. For instance, under 294 and 147 MN, the layer thicknesses became 2.4 and 1.3 nm, respectively. No dislocations were activated, and no faults were formed, even

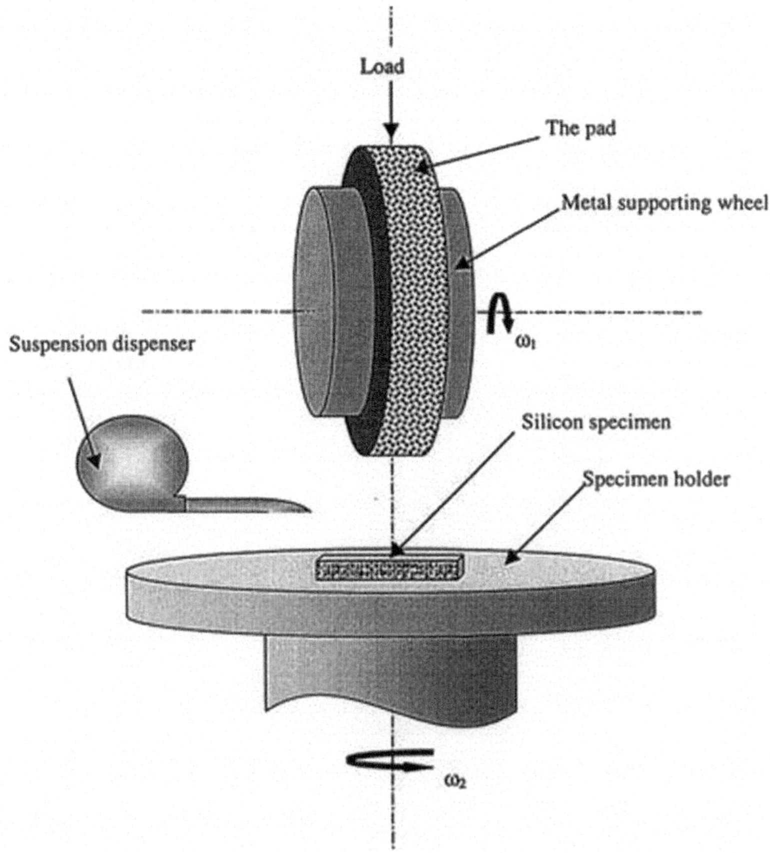

Fig. 12.1 Schematic diagram of the sliding system. L. Zhang and I. Zarudi, Wear, 225, 669 (1999). With kind permission of Elsevier

under the deep scratches were generated shown in Fig. 12.4. The present results show that on the nanometric scale deformation via amorphous phase transformation is a more energetically favorable deformation than that via dislocations. Since the Rms surface roughness of a silicon specimen was only 3 nm, its surface can be considered as absolutely smooth (see Fig. 12.3) compared with its counterpart surface, i.e., the pad–particle surface (see Fig. 12.2), that has a Rms roughness of 60 mm.

The sliding deformation in the system consists of the following two distinct stages under the sliding load:

Stage 1: Micron-scale deformation occurs in the pad only, i.e., no penetration takes place in silicon in this stage.
Stage 2: Nano-scale deformation happens in silicon while the pad deformed in Stage 1 keeps its deformation state unchanged.

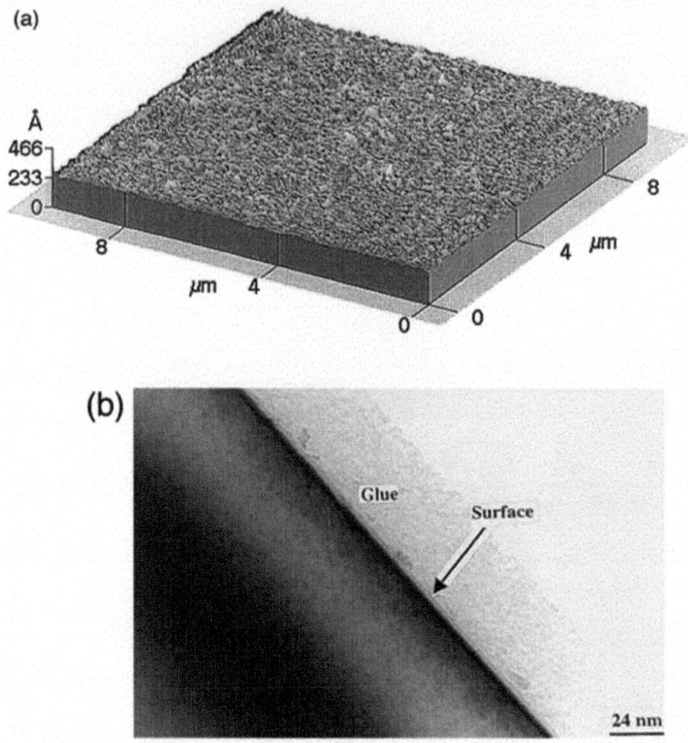

Fig. 12.2 A silicon specimen before sliding test: **a** surface topography; **b** subsurface structure. L. Zhang and I. Zarudi, Wear, 225, 669 (1999). With kind permission of Elsevier

Fig. 12.3 A typical silicon surface after sliding. L. Zhang and I. Zarudi, Wear, 225, 669 (1999). With kind permission of Elsevier

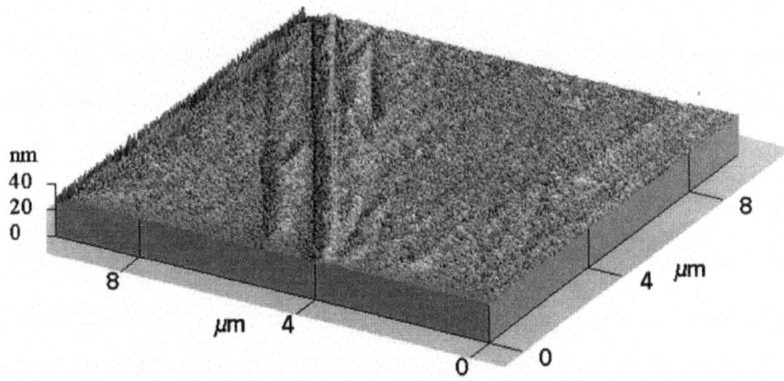

Fig. 12.4 Scratches generated by two-body abrasion. L. Zhang and I. Zarudi, Wear, 225, 669 (1999). With kind permission of Elsevier

In summary, chemical reaction during sliding plays an important role in the nanowear deformation of a mono-crystalline silicon component. The effect is through the change of atomic bonding in the amorphous layer. The increase of the normal sliding load enhances greatly the chemical effect and increases the wear rate. Dislocations do not appear in a mono-crystalline silicon subjected to nano-sliding. Oxygen penetration into the amorphous zone during sliding has been observed and the most probable reason of oxygen penetration into the subsurface is believed to be enhanced by tensile stress field associated with mechanical loading and controlled by the diffusion coefficient of oxygen. Chemical reaction that occurs changes the transformation threshold of transformation from crystalline silicon to amorphous. As would be expected, under smaller load the pad deforms less with less rough surface as a result and vice versa. It is believed that every time an asperity-silicon interaction occurred, it pushed oxygen into the silicon subsurface, accelerated by its tensile stress field generated as mentioned. Since the pushing frequency is larger at the higher load, the thickness of the amorphous layer which is accelerated by chemical reaction is larger. Further investigation is required to understand the rate of penetration of oxygen the origin of threshold change and the effect of wear rate. Oxygen penetrates into the amorphous layer, changes the atomic bonding of silicon, alters the threshold for amorphous transformation and accelerates wear.

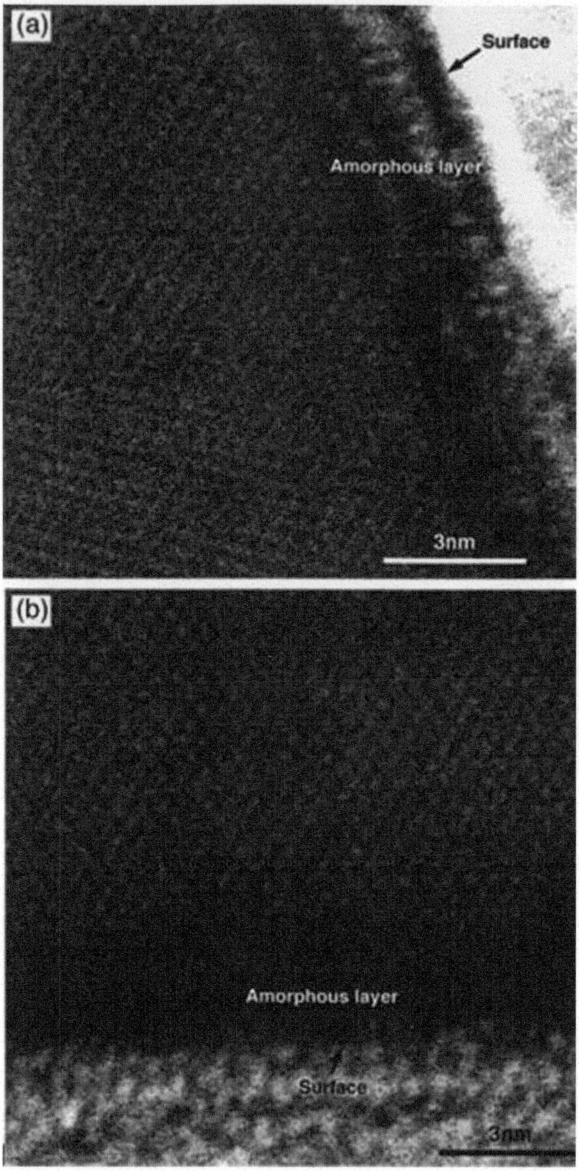

Fig. 12.5 High-resolution images of sub-surfaces under various normal sliding loads, where each spot represents a silicon atom: **a** 441 mN, **b**147 mN. L. Zhang and I. Zarudi, Wear, 225, 669 (1999). With kind permission of Elsevier

12.3 Tensile Deformation in Nanostructure–Effect of Environment

12.3.1 Effect of Oxygen

12.3.1.1 In Si

Elastic, plastic, and fracture properties are key issues in characterizing mechanical behavior of material in nanomaterials as well and silicon is no exception. However, in this section the effect of oxygen on these properties is of interest, due to their application in industry in various technological devices. Silicon nanowires (NWs) are nanostructures used for fabricating various electronic and optoelectronic nanodevices. In situ ultrahigh-resolution TEM (UHRTEM) results of mechanical properties of nanowires is presented below. Plastic deformation was observed at the center of the NW and no abrupt brittle cleavage crack formation was seen. However, instead, the NW transformed from crystalline structure to amorphous at the central region of the axially extended NW. The Si NW samples having 15–70 nm diameter; below 70 nm a large elastic strain was observed, and above it a large plastic strain rate deformation was achieved, which was kept at about 10^{-5} s $^{-1}$. A large reduction in the diameter ratio and in the elongation ratio was obtained at ambient temperature as illustrated in Fig. 12.6. The ratio of initial diameter over breakpoint diameter is given as percentage. Extrapolation of the plot of Fig. 12.6 gives a brittle–ductile transition diameter of around 60 nm for ambient temperature. Above this diameter it is expected that brittle fracture might occur.

The deformation in the Si NW is superplastic and it was observed through an in situ axial tensile experiment using TEM. The images are shown in detail in Fig. 12.7. In this section the interest is the effect of oxygen during deformation of Si NWs. Diffusion into the lattice, especially in the strained region takes place during plastic deformation. Data were collected simultaneously while conducting the axial tensile tests using in situ electron loss spectroscopy (EELS) and TEM. Figure 12.8 shows the results; the EELS spectra are shown in (a) and (b) before and after superplastic deformation. In Fig. 12.8g, h in situ EDS line scans of silicon and elemental oxygen across the Si NW prior to and after necking are spectra are shown in (a) and (b) before and after superplastic deformation. In Fig. 12.8 g, h in situ EDS line scans of silicon and elemental oxygen across the Si NW prior to and after necking are seen. There is an oxygen-rich layer in the oxygen spectra both prior to and after plastic deformation. The oxygen-rich layer is about 3 nm prior to plastic deformation, which is close to the EELS analysis results and the HRTEM observation. With plastic necking, the oxygen layer was reduced to about 2.5 nm, remaining at the same level, as shown in Fig. 12.8h. The thin SiOx layer did not diffuse along the radial direction to contribute to the initiation of dislocation plasticity and large strain plasticity (LSP) of Si NWs.

From the direct atomic-scale imaging of emitted dislocations at ambient temperature in Si NWs during axial tensile experiments, we can correlate the plasticity and LSP of Si NWs with dislocation velocity and the Young's modulus. The plasticity

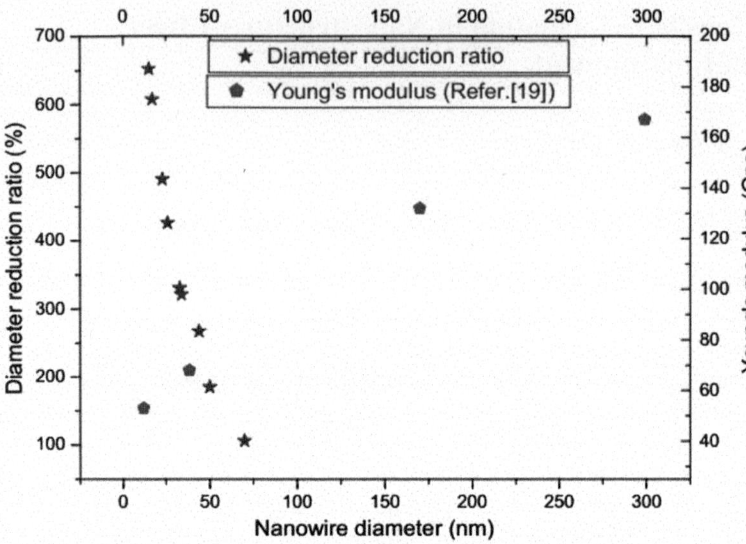

Fig. 12.6 The diameter reduction ratio as a function of nanowire diameter in the case of axial tension at a low temperature close to room temperature. Young's modulus values as a function of diameter are plotted in the same figure for comparison. X. Han, K. Zheng, Y. F. Zhang, X. Zhang, Z. Zhang and Z. L. Wang, Adv. Mater., 19, 2112 (2007). With kind permission of John Wiley and Sons. Ref 19 of the Fig is X. X. Li, T. Ono, Y. L. Wang and M. Esashi, Appl. Phys. Lett., 83, 3081(2003). Noted in the references as Li et al. (2003)

of a material, namely, whether it is brittle or ductile is controlled by the dislocation velocity, which is regarded as the rate of emission of dislocations and/or dislocation motion. It is given by

$$v = A\tau^m \exp\left[\frac{-U}{K_B T}\right] \tag{12.1}$$

where υ is dislocation velocity, A and m are constants, τ is shear stress, k_B is the Boltzmann constant, T is absolute temperature, and U is the activation energy of a dislocation. The activation energy regarded as constant, can be correlated to shear modulus G and elastic modulus E, respectively by

$$U \infty G a_0 \tag{12.2}$$

and

$$U \infty E a_0/[2(1+\gamma)]$$

where a_0 is the lattice constant and γ is the Poisson ratio.

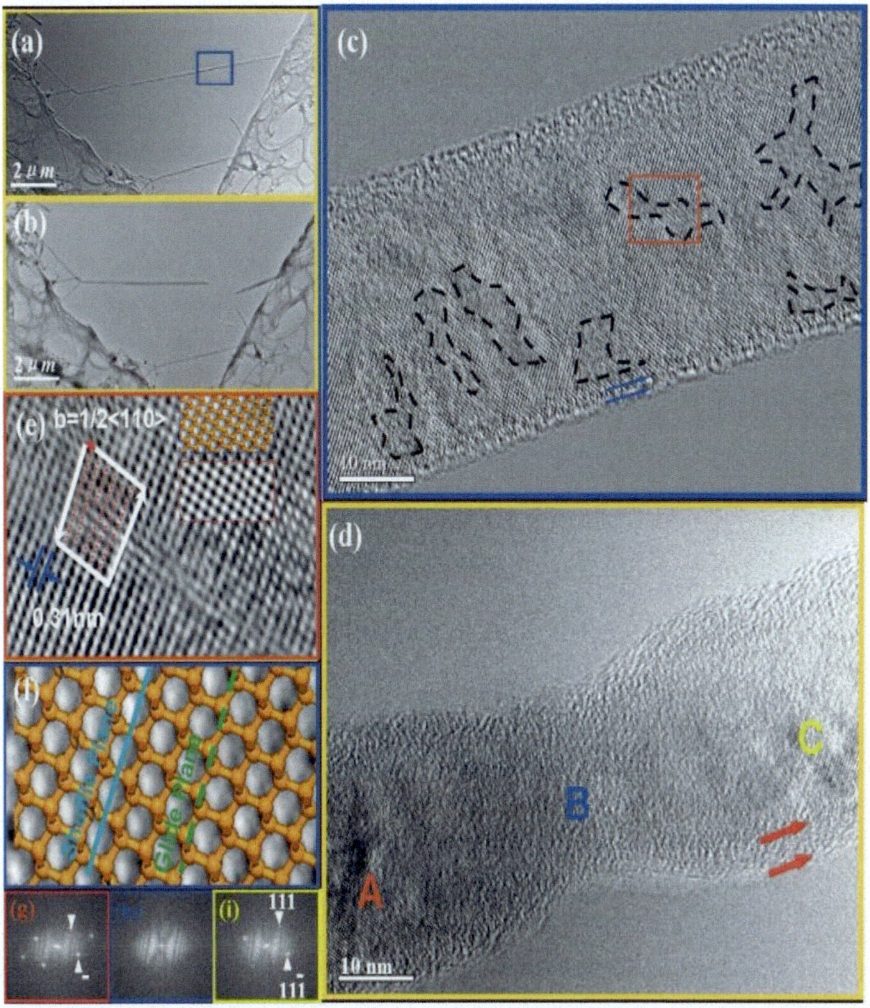

Fig. 12.7 Atomic level in situ high resolution electron microscopy study of an axially extended Si NW. **a**, **b** Low magnification images prior to and after the Si NW is plastically broken. **c** HRTEM image of the Si NW in a status of elastic–plastic transition and prior to severe plastic deformation. **d** The same Si NW after largestrain plastic deformation with a neck in the middle of the NW. **g**, **i** The FFT electron diffraction patterns taken from areas A–C in **c**, respectively. **e** Enlarged HRTEM image taken from the framed region in c, which indicates the initiation features of plasticity with dislocations. The dislocation structure is highlighted by dotted red lines. A Burgers vector circuit is drawn, indicating a perfect edge dislocation. The simulated HRTEM images of the perfect crystalline lattices and the head–tail connection model are shown as insets. **f** Schematic atomic structure of Si atomic lattice projected along the [110] orientation; the shuffle and glide planes are indicated. X. Han, K. Zheng, Y. F. Zhang, X. Zhang, Z. Zhang and Z. L. Wang, Adv. Mater., 19, 2112 (2007). With kind permission of John Wiley and Sons

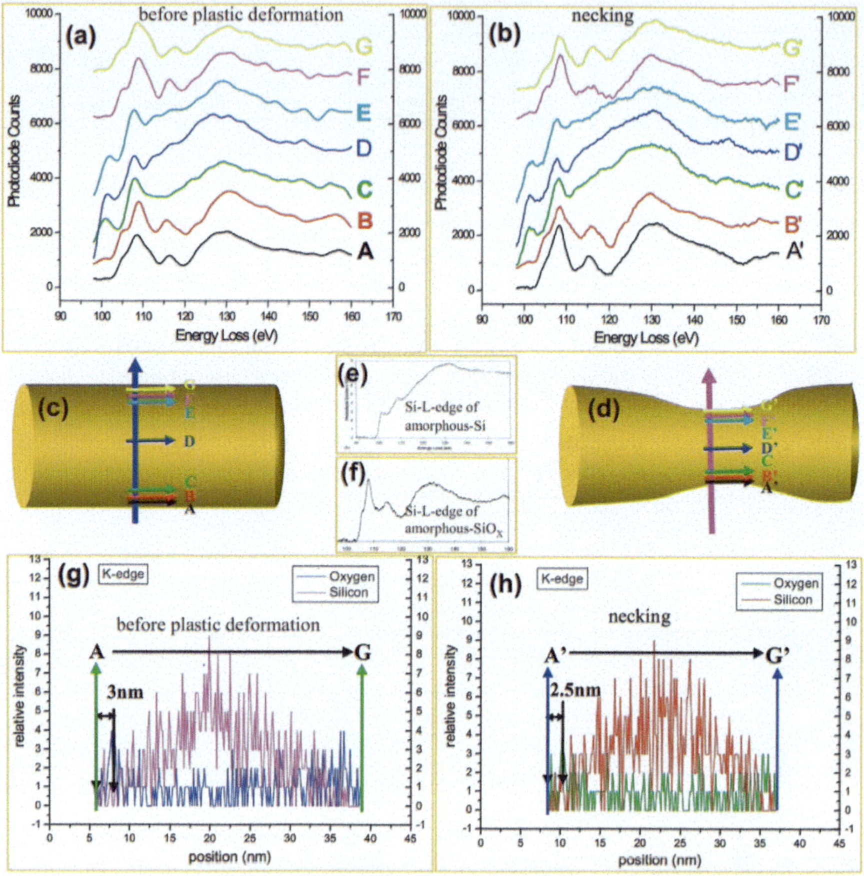

Fig. 12.8 Series of EELS spectra and EDS line scans taken from a Si NW prior to and after large-strain plastic deformation. Both series of EELS spectra were taken across the NW width. **a** A series of spectra taken from the Si NW prior to plastic deformation. **b** A series corresponding to the NW with a superplastic necking feature. **c, d** Schemes of the locations at which the spectra were taken for (a) and (b), respectively. **e, f** Schulmeister et al. (2003) EELS spectra for the Si L edge of Si and amorphous SiO_x, respectively. **g, h** Line scan EDS spectra of elemental oxygen and silicon for the Si NW before and after neck formation. The surface oxygen layer thickness was measured as the distance from the starting position A or A' to the first peak position of oxygen in both oxygen spectra. X. Han, K. Zheng, Y. F. Zhang, X. Zhang, Z. Zhang and Z. L. Wang, Adv. Mater., 19, 2112 (2007). With kind permission of John Wiley and Sons

The elastic modulus of Si is related to the size of the NW, and it decreases with decreasing NW diameter and film thickness. Assuming that a_0 and γ are constants, the relation given below is obtained

$$U_N = \frac{E_N}{E_B} U_B \tag{12.3}$$

where UB and EB are the dislocation activation energy and Young's modulus of bulk silicon, respectively, and U_N and E_N are those of the NW. The dislocation velocity in the NW can thus be obtained as

$$v_N = A\tau \overset{m}{N} \exp\left[\frac{-E_N U_B}{K_B K_B T}\right] \tag{12.4}$$

With $U_B = 2.2$ eV and $E_B = 169$ GPa one obtains for the dislocation activation energy change with diameter and for the velocity for the NWs the relations shown in Figs. 12.9a, b, respectively. It is seen that the activation energy, U_N decreases with decreasing NW diameter and is significantly lower than that of the bulk material. Also, v/v_{av} shows a similar tendency to the diameter reduction ratio. Thus, the diameter reduction ratio of the Si NW is related to the dislocation velocity.

In summary, room-temperature large strain plasticity was directly observed in situ in an ultrahigh-resolution electron microscope in single-crystalline Si NWs. The reduction in NW diameter before fracture occurred, is much higher -about three orders of magnitude higher than that of bulk Si. The ambient temperature Si NW tensile plasticity is initiated by the emergence of a high-density dislocations. Before the Si NW became amorphous plastic formation of long necks occurred, which eventually resulted in fracture. The size-dependent fracture mechanism is attributed to the scalerelated dislocation activities.

12.3.1.2 In SiC

The lack of information on the effect of oxygen environment in pure nanostructure.

SiC, it was felt that it might be of interest to present the subject as Hi-Nicalon woven SiC fibers (2-D woven SiC) comprised of β-SiC nanocrystals with the purpose of evaluating the fracture strength in various environments, among them pure oxygen. It was found that exposure to high oxygen atom concentration in air and oxygen plasmas results in a more severe embrittlement over shorter time scales, corresponding to degradation rates up to 200 times higher than those reported for static heating in conventional furnaces where only molecular oxygen is present, whereas exposure to nitrogen plasma showed little change in fiber morphology. The origin of the embrittlement at microscopic scale is related to surface interactions at high temperatures leading to oxidation and SiO_2 formation. Further microcracks are also formed on the surface. The consequence of such surface reactions is harmful to outer fabric materials for space applications eliminating their proper thermal protection required in atmospheric entry on earth and other planets which are associated with heating.

20-mm diameter coupons were made by mechanically punching a sheet of SiC fabric. The fabric coupons had a bulk thickness of 0.5 mm and an average mass of 0.14 g.

The aerothermal testing system of the 2-D woven SiC fibers is seen in Fig. 12.10. Constant surface temperature (within 25 °C) was attained in air-nitrogen plasmas

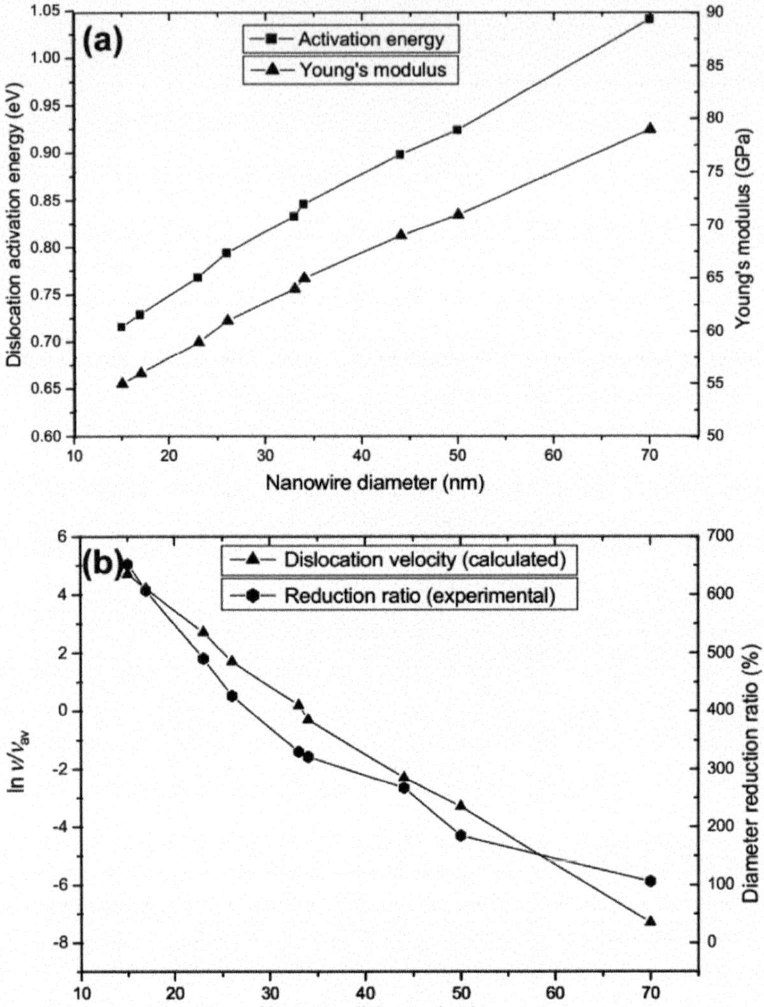

Fig. 12.9 **a** The dislocation activation energy of nanowires UN with respect to the diameter of the nanowires. The extrapolated Si NW Young's modulus is also plotted. **b** The normalized dislocation velocity ratio v/v_{av} as a function of the diameter of the Si nanowires. The diameter reduction ratio is plotted in the same figure for comparison. X. Han, K. Zheng, Y. F. Zhang, X. Zhang, Z. Zhang and Z. L. Wang, Adv. Mater., 19, 2112 (2007). With kind permission of John Wiley and Sons

after a short time of 5 s, while in pure nitrogen it required exposure up to 1 m before achieving constant surface temperature. The surface temperature for tests at different plasma gases is illustrated in Fig. 12.11. Complete details of the experimental system and the procedure are described in Owens et al. (2015) and are summarized in their Table 12.1.

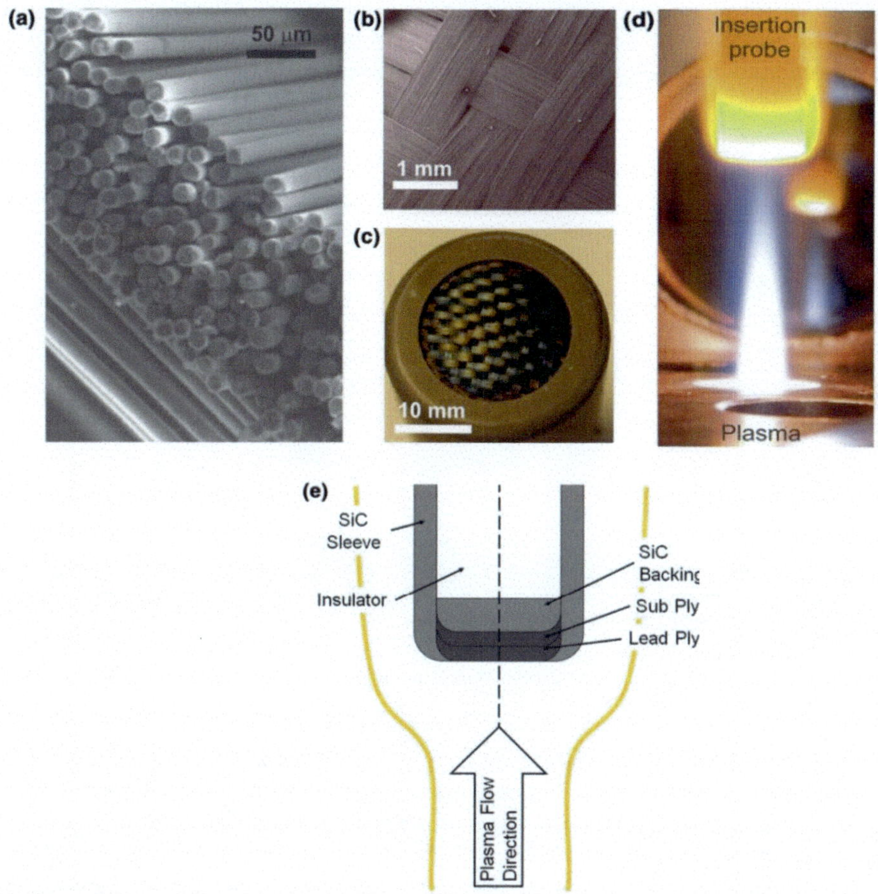

Fig. 12.10 Aerothermal testing of 2-D woven SiC fibers. **a** Cross-sectional and **b** top views of woven SiC fibers by SEM. **c** Woven Hi-Nicalon SiC lead-ply coupon in SiC sleeve. **d** Probe insertion in the 30 kW inductively coupled plasma torch facility. **e** Schematic illustration of a longitudinal cross-section view of the insertion probe. W. Owens, D. Merkel, F. Sansoz and D. Fletcher, J. Am. Ceram. Soc., 98, 4003 (2015). With kind permission of John Wiley and Sons

Tensile tests at room temperature were performed on coupons following plasma exposures of 30, 60, 120, and 240 s. The load vs. displacement is seen in Fig. 12.12.

Tensile deformation was carried out at a constant displacement rate of 1 mm/min until fracture occurred as seen in Fig. 12.12b. The load–displacement curve of Fig. 12.12a shows that the cloth deformation is quasi-linear with a clean break at peak load F_{max}. The fracture strength s_f is estimated by using

$$\sigma_f = \frac{F_{max}}{\pi f (D/2)^2} \tag{12.5}$$

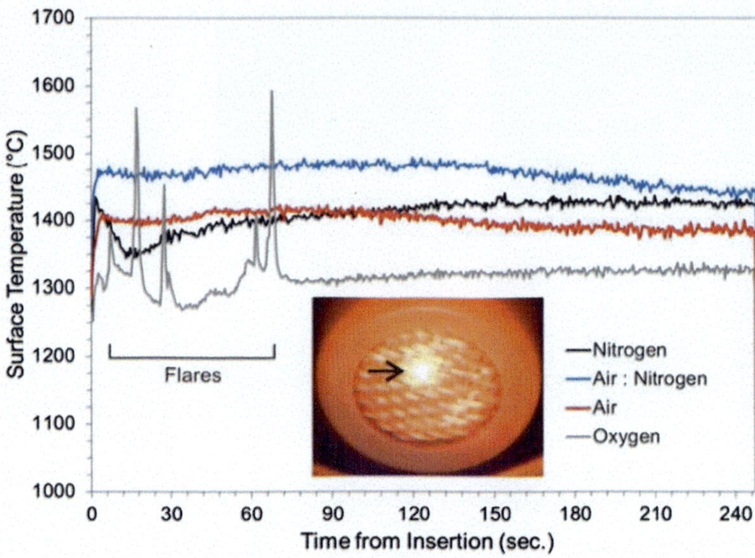

Fig. 12.11 Two-color pyrometer measurements of coupon surface temperature for tests in different plasma gases. Inset shows a snapshot taken from an in-situ video recording device during a surface flare in oxygen plasmas. W. Owens, D. Merkel, F. Sansoz and D. Fletcher, J. Am. Ceram. Soc., 98, 4003 (2015). With kind permission of John Wiley and Sons

Here f is the total number of failed fibers, and D is the average fiber diameter. f was determined by counting the total number of broken tows times the average number of fibers per tow. Using SEM, it was found that D ~14 μm and an average of 500 fibers per tow.

The structure of 2-D woven SiC fibers after exposure to pure nitrogen plasma is illustrated I Fig. 12.13. No significant change (such as discoloration) at the surface is observed after exposure to nitrogen plasma as indicated by Fig. 12.13a. Further, SEM image of the surface after exposure to nitrogen plasma remained smooth as seen in Fig. 12.13b. A few fibers showed localized surface damage in the form of pitting and fiber thinning, as indicated by arrows in Fig. 12.13b. In these damage locations EDS indicates large nitrogen and oxygen concentrations which shows possible formation of stable Si_2N_2O phase. The fracture strength decreases after seconds of exposure to nitrogen plasmas seen in Fig. 12.14 for both the lead-ply and sub-ply. The fracture strength decreases after a few seconds exposure, but it levels of at ~50% strength after 30 s exposure to nitrogen. but retain their strength after longer than 30 s exposure time.

After exposure to pure oxygen plasma a large change is indicated as shown in Fig. 12.15. SiO_2 is formed to about 2 nm thick in pure oxygen plasma already after 60 s at temperatures above 1000 °C as seen in Fig. 12.15b, c. Compare the fracture strength reduction in nitrogen atmosphere (Fig. 12.14) with fiber oxidation in both air

Table 12.1 Comparison of properties of three generations of SiC fibers. J. Donge, W. Bai, G. Huang, L. Han and G. Zhang, Earth and Environmental Sci., 571, 012,147 (2020). Open access

	Name	Manufacturer	Thermal expansion coefficient Below 1.000 / $(10^6 \cdot C^{-1})$	Axial thermal conductivity at room temperature /(W.m.K^{-1})	Tensile strength at room temperature /(Gpa)	Young modulus at room temperature /(Gpa)
First generation	Nicalon200	Japanese carbon company	3.2	3.0	3.0	200
	Tyranno LOX-M	Japanese UBE Industries Ltd	3.1	1.5	3.3	285
Second generation	Hi-Nicalon	Japanese carbon company	3.5	8.0	2.8	270
	Tyranno LOX-E	Japanese UBE Industries Ltd	N/A	N/A	2.9	200
	Tyranno ZM	Japanese UBE Industries Ltd	N/A	2.5	3.4	200
	Tyranno ZE	Japanese UBE Industries Ltd	N/A	N/A	3.5	233
Third generation	Tyranno SA 1	Japanese UBE Industries Ltd	N/A	6.5	2.8	375
	Tyranno SA 3	Japanese UBE Industries Ltd	N/A	6.5	2.9	375
	Sylramic	American dow corning	5.4	4.6	3.2	400
	Sylramic iBN	American dow corning	5.4	> 4.6	3.5	400
	Hi-Nicalon Type-S	Japanese carbon company	N/A	1.8	2.5	400

and oxygen plasmas which leads to more severe strength reduction and embrittlement than nitrogen plasmas and shown in Fig. 12.16 for different oxidizing conditions.

The effect of surface temperature variation on strength reduction and embrittlement of 2-D woven SiC fibers is illustrated in Fig. 12.17. They were exposed to dissociated oxygen, air and 100% nitrogen plasmas producing high surface temperatures. Also tensile testing at room temperature were investigated to evaluate fracture

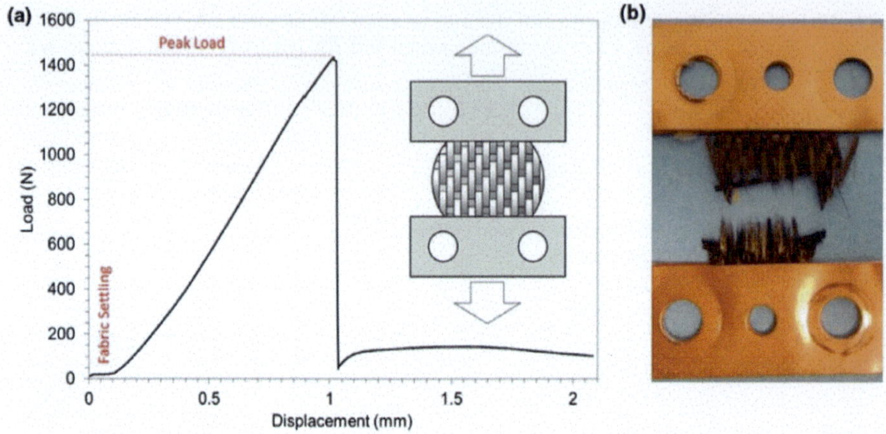

Fig. 12.12 Tensile testing of 2-D woven SiC fibers. **a** Representative loaddisplacement curve showing a clean break at peak load. Inset shows the yarn orientation with respect to the loading direction. **b** Coupon after fracture. W. Owens, D. Merkel, F. Sansoz and D. Fletcher, J. Am. Ceram. Soc., 98, 4003 (2015). With kind permission of John Wiley and Sons

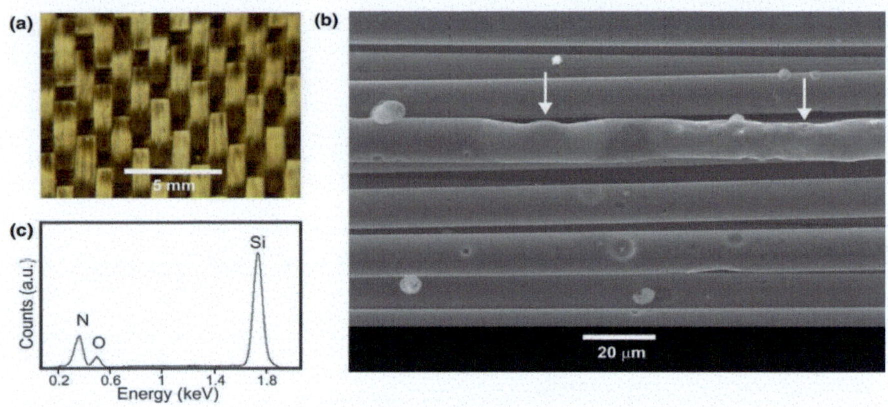

Fig. 12.13 Structure of 2-D woven SiC fibers after pure nitrogen plasma exposure for 240 s. **a** Optical micrograph and **b** SEM image of the exposed lead-ply surface. **c** EDS results from damaged locations indicated by arrows. W. Owens, D. Merkel, F. Sansoz and D. Fletcher, J. Am. Ceram. Soc., 98, 4003 (2015). With kind permission of John Wiley and Sons

strength reduction and embrittlement in 2-D woven SiC fibers. Exposure to high oxygen atom concentration in air and oxygen plasmas results in a more severe embrittlement over shorter time scales, corresponding to degradation rates up to 200 times higher than those reported for static heating in conventional furnaces where only molecular oxygen is present. Strength reduction by exposure to nitrogen resulted only 50% strength reduction compared to the strength of virgin fiber. The origin of the accelerated embrittlement in oxidizing plasmas was found to be associated with

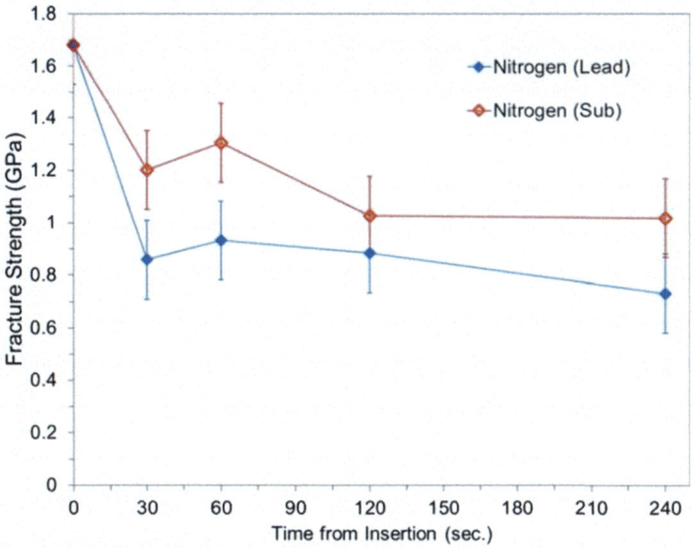

Fig. 12.14 Room-temperature fracture strength of the lead-ply and sub-ply coupons after exposure to nitrogen plasma. The virgin material strength before exposure or time zero was averaged over 3 tests. W. Owens, D. Merkel, F. Sansoz and D. Fletcher, J. Am. Ceram. Soc., 98, 4003 (2015). With kind permission of John Wiley and Sons

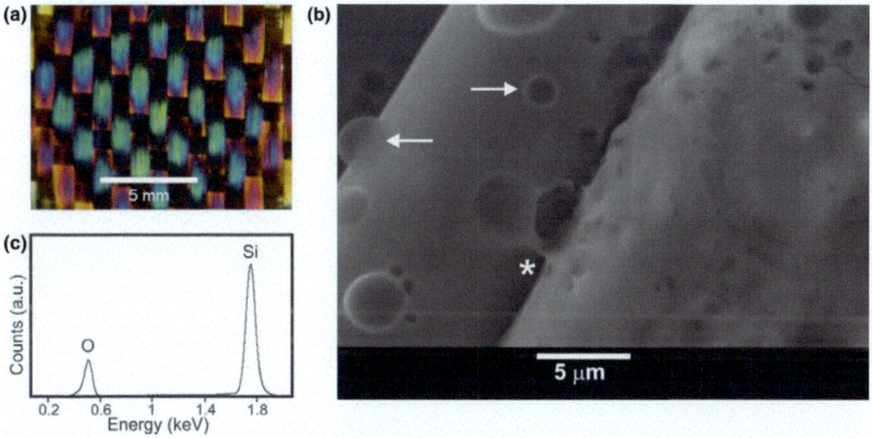

Fig. 12.15 Structure of 2-D woven SiC fabric after pure oxygen plasma exposure for 240 s. **a** Optical micrograph and **b** SEM image of the exposed lead-ply surface with arrows showing the formation of gas bubble at high temperature and a star for an example of surface bonding between two fibers. (c) EDS spectrum of the silica surface. W. Owens, D. Merkel, F. Sansoz and D. Fletcher, J. Am. Ceram. Soc., 98, 4003 (2015). With kind permission of John Wiley and Sons

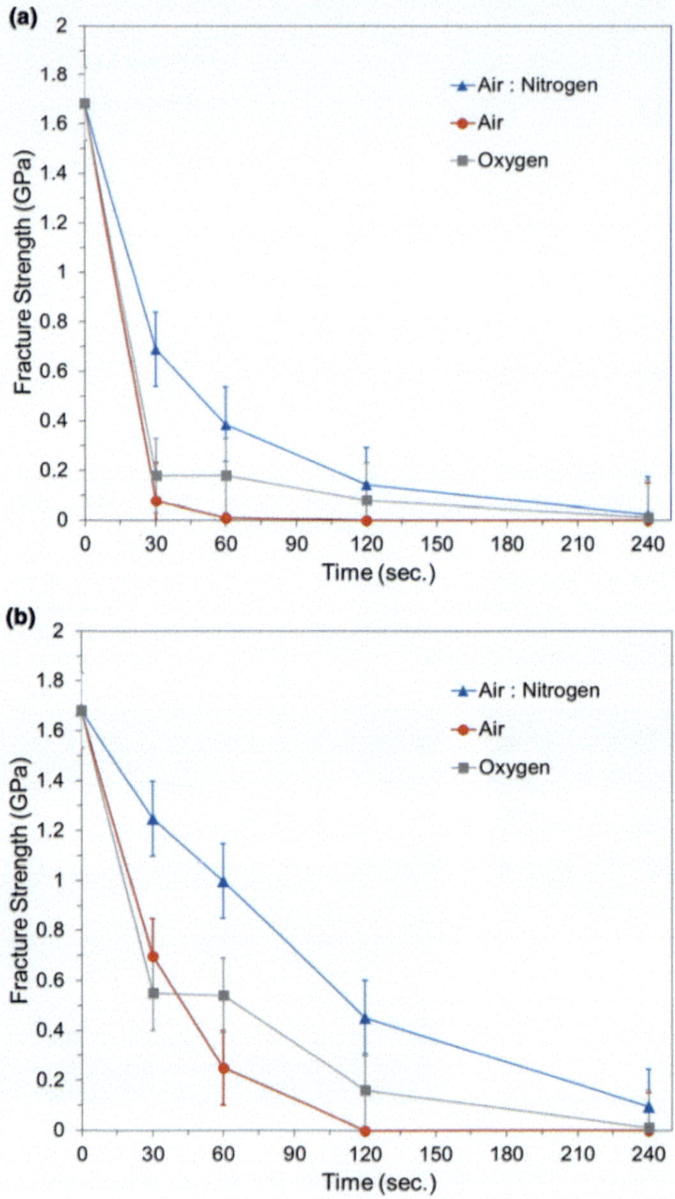

Fig. 12.16 Fracture strength after exposure to different oxidizing conditions for **a** lead-ply and **b** sub-ply coupons. W. Owens, D. Merkel, F. Sansoz and D. Fletcher, J. Am. Ceram. Soc., 98, 4003 (2015). With kind permission of John Wiley and Sons

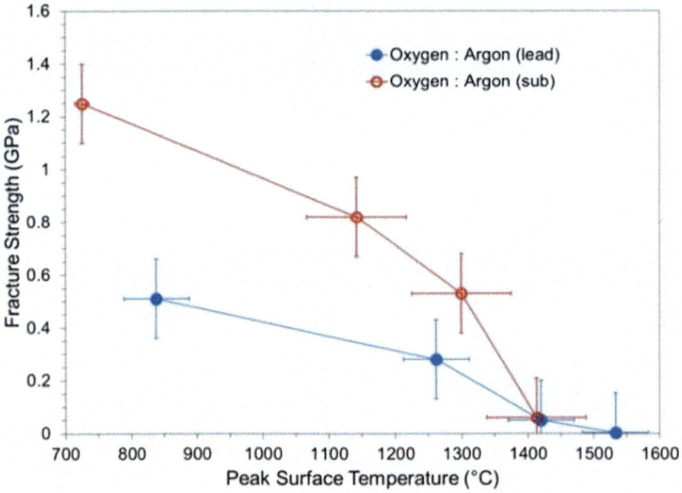

Fig. 12.17 Effect of surface temperature on fracture strength for lead-ply and sub-ply coupons exposed to different oxygen-argon plasma mixtures for 120 s. W. Owens, D. Merkel, F. Sansoz and D. Fletcher, J. Am. Ceram. Soc., 98, 4003 (2015). With kind permission of John Wiley and Sons

the formation of a viscous silica surface layer leading to the development of gas bubbles, inter-filament adhesion, and critical flaws. Further, The fracture resistance of 2-D woven SiC fibers is totally lost after 120 s of plasma exposure at 1400 °C.

In summary, fracture strength reduction and embrittlement in 2-D woven SiC fibers were considered in 100% nitrogen plasmas producing high surface temperatures. Little change in fiber morphology led to moderate strength reduction of ~50% within 30 s. However, exposure to high oxygen atom concentration results in severe embrittlement after a short time. These results are associated with formation of a viscous silica surface layer leading to the development of gas bubbles, inter-filament adhesion, and critical flaws. The fracture resistance of 2-D woven SiC fibers is totally lost after 120 s of plasma exposure at 1400 °C.

12.3.2 Effect of Moisture

12.3.2.1 In SiC

Fragments of a short note on the extrusion of SiC ceramics, the influence of moisture content -apparently by X. Dai and coworkers-was mentioned and illustrated by two figures, reproduced below with very limited information (Figs. 12.18 and 12.19).

More information on environmental effects on SiC/SiC composite, intended for nuclear reactor systems use, among them also the effect of moisture is mentioned.

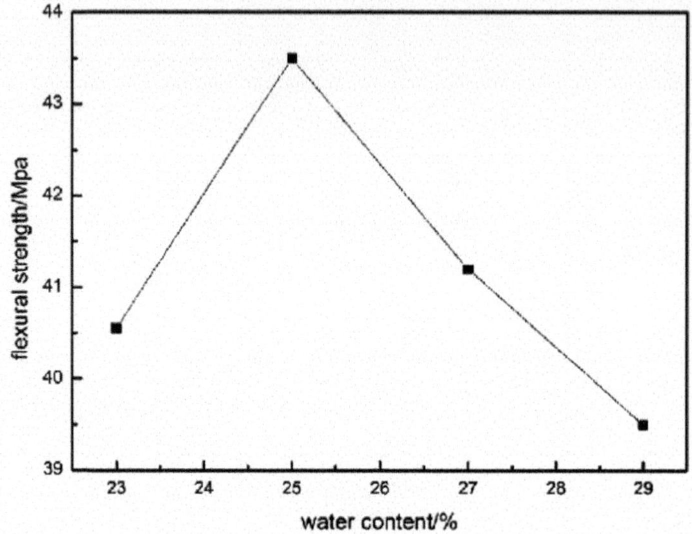

Fig. 12.18 Effect of water content on flexural strength, X. Dai, Fan Shen, Jia-you Ji, Shu-Ling Wang and Man Xu, Maater. Sci., DeStech Trans. Eng. Techno. RTes. (2018). Open access

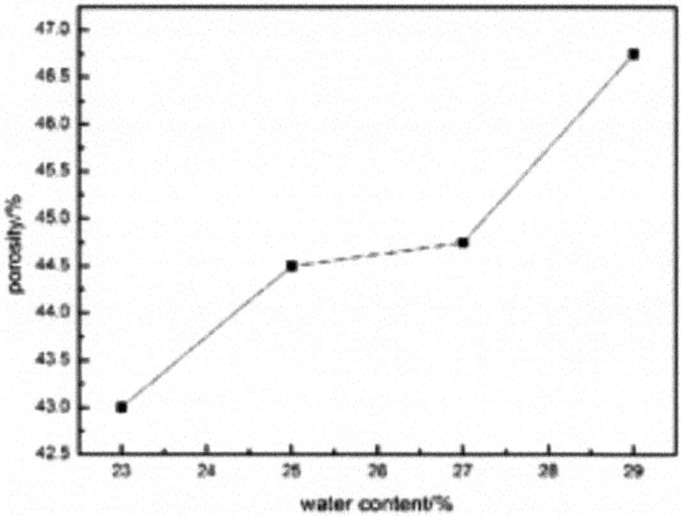

Fig. 12.19 Effect of water content on porosity. X. Dai, Fan Shen, Jia-you Ji, ShuLing Wang and Man Xu, Maater. Sci., DeStech Trans. Eng. Techno. RTes. (2018). Open access

Recall-as already stated along this book- that SiC is usually used in a composite form with SiC fibers (whiskers) or together with other constituents and only seldom it is mentioned for technological purposes as a free pure component.

Water vapor induces passive mode of oxidation according to the equations

$$SiC_{(s)} + 2H_2O_{(g)} \rightarrow SiO_{2(s)} + CH_{4(g)} \text{at } T < 1400\ ^{o}C \tag{12.6}$$

$$SiC_{(s)} + 3H_2O_{(g)} \rightarrow SiO_{2(g)} + CO_{2(g)} + 3H_{2(g)} \text{at } T > 1400\ ^{o}C \tag{12.7}$$

The rate of passive oxidation is enhanced by the presence of water vapor in oxygen or air. The rate enhancement is explained by water having higher affinity to silica than oxygen and thus higher permeability through the scale SiO_2 layer. Further water can break the Si–O bonds, which enhances diffusion. In the presence of high P_{H2} gasification may occur through reduction of the silica layer, according to

$$SiO_2(s) + H_2(g) = SiO(g) + CH_4(g) \tag{12.8}$$

The effect of water on SiC/SiC composites is of great interest since its wide range use for example in gas turbines where they are exposed to high temperature steam. Oxidation by steam as indicated by the chemical reaction earlier, forms SiO_2 layer. In the presence of H_2O, however, this layer no longer protects the underlying SiC fiber and gasification continues (as seen in the relations above) by diffusion through a porous boundary layer.

For additional information on the possible chemical reactions, on passive oxidation and also on active oxidation of SiC in the presence of water in the system one is referred to the published work Dai et al. (2018)

12.4 Creep in Nanomaterials Effect of Environment

12.4.1 Effect of Oxygen

12.4.1.1 In SiC

It has been indicted (for example by NASA Barttolotta n d) that the non-oxide SiC/SiC ceramic matrix composite (CMC) has the most promise for high temperature, high oxidation applications. Its disadvantage is its tendency towards embrittlement which must be overcome. The embrittlement of SiC/SiC CMCs is an oxidation embrittlement of the fiber/matrix interphase occurring on exposure to a harsh environment.

Seldom one finds in the open literature the combined effect of structural properties and environmental effects. Generally, in publications each of these features appear separately and are considered as such.

SiC often is being used as composites among them as SiC_f/SiC where SiC_f stands for SiC fiber referred to Nicalon fiber and plays an important role in the overall properties in particular the mechanical behavior of the composite. Particular attention has been devoted to developing and improve production methods of SiC fibers, due to their important role in the overall SiC composite properties. Thus, in the progress of developing SiC fibers several generations were produced Dong et al. (2020), characterized by various amounts of oxygen, not deliberately added but rather present as a consequence of production by development of polymer derived SiC fibers.

Nicalon fiber, the first generation SiC fiber, has relatively high oxygen content (13%), and a large number of thermal decompositions occurs in SiC_xO_y phase, resulting in the porous structure, and then oxygen diffuses easily to promote active oxidation.

The second generation SiC fiber has specially taken some measures to reduce the oxygen mass fraction of the fiber and improve the high temperature property of the fiber. The oxygen content of the fiber is reduced to be 0.5%, which determines that SiC_xO_y has less decomposition and compact structure, and is not conducive to oxygen diffusion, so that the degree of active oxidation is reduced. Moreover, the mass fraction of free carbon is relatively high, the preferential oxidation of free carbon has the effect of inhibiting the active oxidation of SiC, the grain size is larger than that of the first generation, the heat resistance and creep resistance are improved to a certain extent. The high temperature resistance and creep resistance of the second-generation fiber are improved, but the oxidation resistance is still not ideal. CMC, which is applied in many fields such as aviation and aerospace, requires its reinforced fibers to be used stably in air not lower than 1,300 °C and inert atmosphere not lower than 1,600 °C. Therefore, it is necessary to further upgrade SiC fiber and the polycrystalline SiC fiber with close chemical ratio has been developed. In the third generation SiC, therefore, the fiber consists of submicron β-SiC grains, a small amount of carbon and trace oxygen ultimately. This fiber has high Young modulus, high creep resistance, high oxidation resistance and good thermal stability below 1,600 °C. Table 12.1 summarizes the properties of the various three generations SiC fibers. Note the room temperature mechanical properties of these SiC fibers.

In summary, SiC crystal is a covalent compound. In the crystal, carbon and silicon are arranged in sp Han et al. (2007) hybridization to form SiC crystal with diamond structure.

Therefore, the ideally β-SiC can withstand high temperature up to 2,600 °C. However, SiC fiber produced by polymer-derived conversion method is much below this ideal level. Since oxygen in SiC is detrimental, improving the production technology to minimize or avoid the introduction of oxygen during production is essential. At present, the properties of the third generation SiC fiber have met the needs of practical application. It is not only applied in heat resistant parts of aviation and aerospace engines, thermal protection material systems of reusable vehicles etc. but also have application potential in nuclear energy, high-speed brakes, hot-end components of gas turbines, high temperature gas filtration, heat exchangers, etc.

12.4.2 *Effect of Moisture*

12.4.2.1 In Si

The use of silicon is widespread in many micro- and nano-scale technological applications, most notably microelectronics and microelectromechanical systems (MEMS). At room temperature, both single crystal silicon (SCS) and polysilicon are brittle and thus the strengths of such micro- and nano-scale components are brittle fracture strengths, i.e., all deformation is elastic prior to and during fracture. Of the many factors affecting the strength, namely the mechanical properties, here the testing environmental effect, more specifically the humidity is considered.

The mechanical behavior of both, that of SCS and polysilicon are degraded by the environmental conditions, such as humidity; while the degradation of SCS components is slow and may be related to the slow oxidation rate of Si surfaces, the marked effect of humidity on reducing the fatigue resistance of polysilicon should be noted. This degrading effect of humidity has been attributed to a "reaction layer" process by which the surface oxide is thickened by cyclic loading. Generally speaking, the moisture induced subcritical cracking of the surface oxide in polysilicon has also been attributed as the cause of high-cycle fatigue failure of polysilicon microsystems. Further, when either SCS or polysilicon is heated up above 800 °C in inert environments or vacuum, the presence of heterogeneous pitting ruins devices such as thermal actuators within a very short time.

12.4.2.2 In SiC

SiC in particular as a SiC/SiC composite is being considered for use at high temperature and high radiation in various reactors Contescu and Oak (2013). Due to many favorable properties of SiC/SiC including its mechanical properties, information on its chemical reactivity is of prime interest. Under normal operating conditions of helium cooled nuclear reactor systems, even the ultra-high purity helium will have impurities such as water, oxygen, etc. which has to be chemically controlled.

The attractive mechanical properties of SiC/SiC composites (tensile strength, elastic modulus, creep resistance) originate from the reinforcing fibers. Bonding of fibers to the matrix is ensured by a thin layer of compliant material, generally known as the interphase, which enhances the non-brittle mechanical behavior. Pyro carbon is commonly used as the interphase. The properties of SiC matrix depend on the fabrication route. Most common routes, compatible with the nuclear environment, are chemical vapor impregnation (CVI) and nano-infiltration and transient eutectic-phase process (NITE). Both these methods produce highly crystalline, stoichiometric matrices and can be fabricated in a variety of shapes. The CVI process is a variety of the chemical vapor deposition process (CVD) of pyrolytic SiC, which allows deposition of high purity crystalline β-SiC phase. At high temperatures > 2500 °C SiC decomposes according to

$$SiC_{(s)} \rightarrow C_{(s)} + Si(C)_{(l)} \tag{12.9}$$

The coolant helium gas in high temperature reactors is expected to contain low level impurities such as H_2O, O_2 and CO_2 that may cause slow gasification over the lifetime of components. These impurities have oxidation capabilities, that may affect components in reactivity control system, such as the control rods and shutdown rods. The character of oxidation of the ceramic composite constituents (i.e., the various SiC components) depends on the supply of oxygen. With sufficient oxygen supply the formation of a passive protecting SiO_2 (silica) surface layer is formed (termed passive oxidation), but with limited oxidant supply severe corrosion occurs trough volatilization and release of SiO gas phase (termed active oxidation). The reactivity of various components in SiC/SiC composites is presented below.

(i) **Passive Oxidation of Massive SiC**

Oxidation of SiC in air at high temperatures produces a thin silica film by the reaction of:

$$SiC_{(s)} + 3/2\ O_{2(g)} = SiO_{2(s)} + Co_{(g)} \tag{12.10}$$

$$SiC_{(s)} + 2O_{2(g)} = SiO_{2(s)} + CO_{2(g)} \tag{12.11}$$

Oxidation rates of crystalline α-SiC are limited by the transfer of oxygen through the scale. Active oxidation at low oxygen concentration the reaction is

$$SiC_{(s)} + O_{2(g)} = SiO_{(g)} + CO_{(g)} \tag{12.12}$$

(ii) **The passive mode of oxidation by water vapor**

It occurs through the following reactions:

$$SiC_{(s)} + 2H_2O_{(g)} \rightarrow SiO_{2(s)} + CH_{4(g)} \quad \text{at } T < 1400\ ^\circ C \tag{12.13}$$

$$SiC_{(s)} + 3H_2O_{(g)} \rightarrow SiO_{2(s)} + CO_{2(g)} + 3H_{2(g)} \text{at } T > 1400\ ^\circ C \tag{12.14}$$

The presence of water vapor in oxygen or air enhances the rate of passive oxidation, from about 2–3 times. (when 10% H_2O is added to oxygen the enhancement is about 10–20 times (when comparing oxidation in steam versus dry oxygen). The probable reason is the higher affinity of water to silica than oxygen. The higher the moisture content the more important is the weight loss of the passive SiO_2 layer. Active oxidation at low oxygen partial pressure may also occur in the presence of CO_2/CO mixtures. Oxidation occurs of the SiC matrix, of the SiC fibers and of the interface, namely of the entire composite.In gas turbine applications where SiC/SiCs are exposed to high temperature steam, oxidation of SiC by steam according to reactions 12.13 and 12.14 forms a SiO_2 layer. In the presence of H_2O or water rich atmosphere (steam), however, this layer no longer protects the underlying SiC fiber and gasification continues according to for which the rate is proportional with P_{H2O} and $(P_{H2O})^2$, respectively, controlled by the diffusion of volatile species Si-O_x-$H_{y(g)}$ through a porous boundary layer. This is illustrated by the electron microscopy images presented in Fig. 12.20, which also compare the surface of α – and β-SiC after exposure to high pressure steam in air, or dry air only.

$$SiC_{(s)} + H_2O_{(g)} \rightarrow SiO(OH)_{2(g)} \quad (12.15)$$

$$SiC_{(s)} + 2H_2O_{(g)} \rightarrow SiO(OH)_{4(g)} \quad (12.16)$$

In summary, the chemical stability of SiC/SiC composite depends on temperature, exposure time and environment. The most advanced SiC/SiC composite, namely

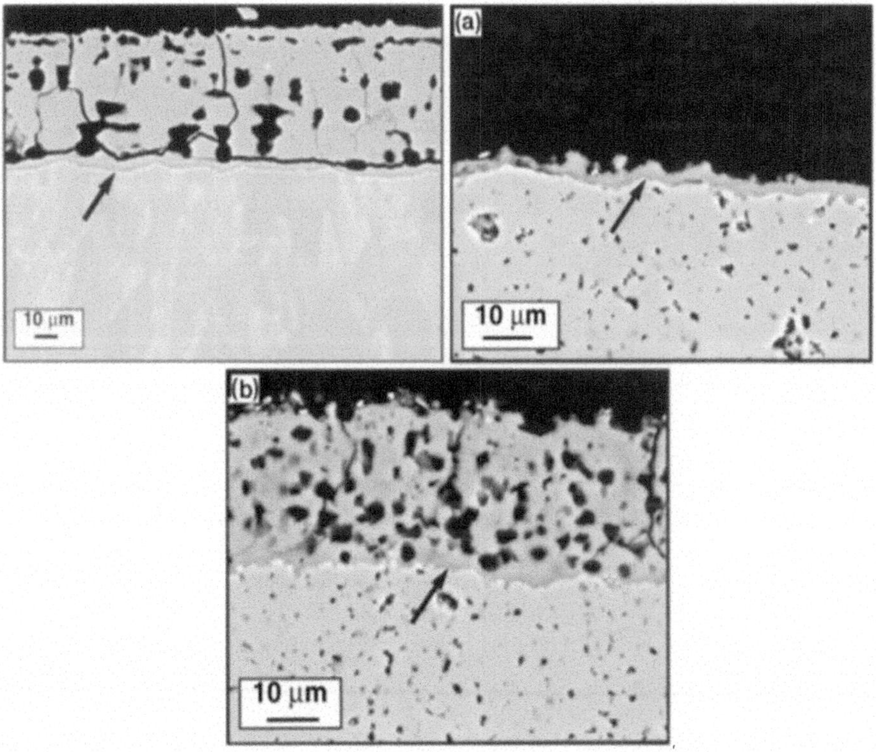

Fig. 12.20 Left-Typical silica scale morphology observed for CVD b-SiC after 500 h of exposure to 10 bar air + H_2O at 1200 °C; right- silica scale on sintered β-SiC after 100 h of exposure to 10 bar of (a) air and (b) air + H_2O at 1200 °C. The arrows show the dense silica layer. Only this layer was formed when H_2O was not present. C. I. Contescu, OAK RIDGE NATIONAL LABORATORY Oak Ridge, Tennessee 37,831–6283. (2013). Unlimited distribution

the third-generation nuclear grade composite, allows them to maintain mechanical properties stability including time dependent and cyclic deformations at the highest temperatures of 1200–1300 °C or even higher in harsh environment Contescu and Oak (2013).

12.4.2.3 In GaAs

During device operation at room temperature changes in the mechanical properties' behavior such as creep and fatigue is probable; the magnitude of these changes is of significant importance. Below the creep behavior and its effect on properties such as hardness and Young's modulus is considered by using nanoindentation technique. The experiments were performed by a depth-sensing indentation system, (Triboscope Hysitron, USA) on a 2 × 2 μm^2 area of (100) GaAs at room temperature. A diamond indenter with a Berkovich tip having a tip radius of 100 nm was applied with load ranges of 500–3000 μN, holding times of 0–60 s at a loading rate of 500 μN/s during a single indentation cycle.

Let us rewrite some of the earlier relations related to the hardness and Young's modulus as

$$H = \frac{p_{\max}}{A} \tag{12.17}$$

$$A = 24.5h_c^2 \tag{12.18}$$

$$E_j = (1 - v_j^2)\left[\frac{1}{E*} - \frac{1 - v_i^2}{E_i}\right]^{-1} \tag{12.19}$$

$$E^* = \frac{\sqrt{\pi}}{2}\frac{\beta s}{\sqrt{A}} \tag{12.20}$$

where P_{max} is the indentation load, A is the contact area, h_c is the contact depth, E_j is the Young's modulus of the thin film, v is Poisson's ratio, E is the Young modulus and *S* is the stiffness of the test material respectively. The subscripts *i* and j refer to the indenter and test material. The indenter properties used in the calculations are $E_i = 1140$ GPa and $v_i = 0.07$ and it was assumed that $v_j = 0.3$ for GaAs. S can be obtained from the initial unloading slope by evaluating the maximum load and maximum depth, which is $S = dP/dh$. The loading–unloading curve can be obtained by using a Berkovich indenter, β is a shape constant that depends on the geometry of the indenter, which is 1.034 for a Berkovich tip.

The effects of creep can be inferred from Fig. 12.21. It illustrates loadingunloading curves for GaAs at a load of 2000 μN and a loading rate of 500 μN/s under hold times of 0, 15, 30 and 60 s. Increasing the hold times increased the maximum penetration depth. The maximum penetration of GaAs is deeper than that observed in Si. Young's modulus and the hardness for GaAs under different hold times, at a maximum load of 2000 μN and at a loading rate of 500 μN/s are shown in Fig. 12.22a, b respectively. From these figures, it can be seen that Young's modulus and the hardness for GaAs decreased as the hold times were increased until shortly after 40 s when they kept almost constant. This indicates that shortly after the hold time of 40 s the creeping effect minimized.

Young's modulus for GaAs is ~87 GPa. The hardness for GaAs is ~7.2 GPa. Note that the Young's modulus and the hardness are compared to Si. Further, it might be of interest to show that the plastic energy and the ratio of the elastic recovery are affected by the hold time at a load of 2000 μN and a loading rate of 500 μN/s. Increasing the hold time increased the plastic energy and the ratio of elastic recovery as seen in Fig. 12.23. Both the plastic energy and the ratio of elastic recovery gradually reached a constant value shortly after at the hold time of 40 s.

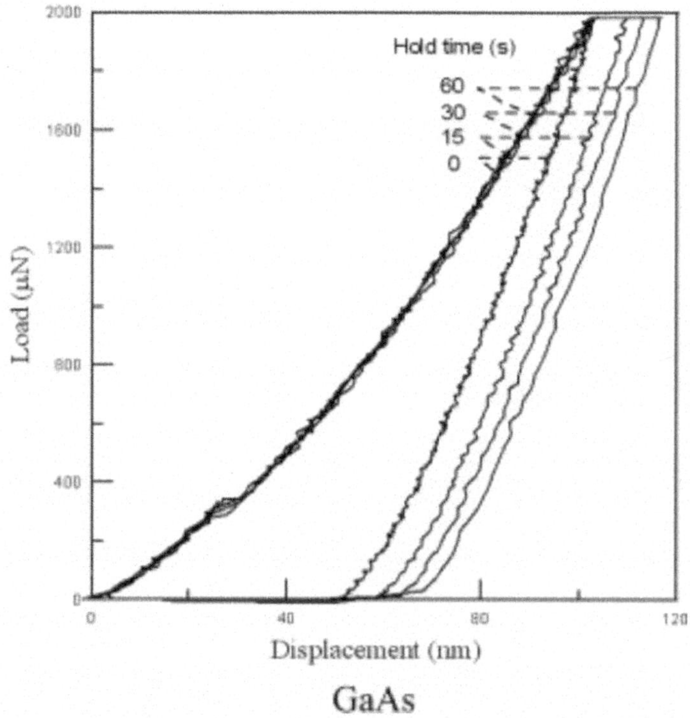

Fig. 12.21 The loading–unloading curves for GaAs at the maximum load of 2000 μN and the loading rate of 500 μN/s under different hold times. Te-Hua Fang, Win-Jin Chang and Chao-Ming Lin, Microelectronic Eng., 77, 389 (2005). With kind permission of Elsevier

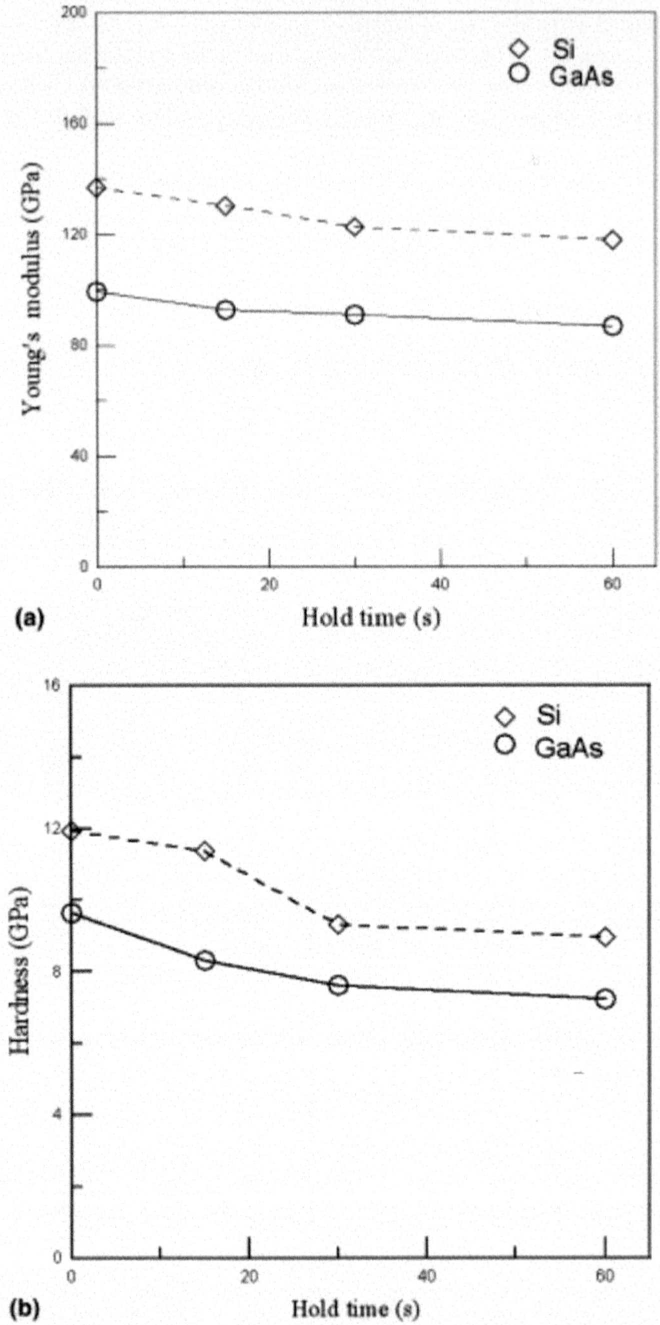

Fig. 12.22 The relationship between the mechanical properties and the hold time at the load of 2000 μN and the loading rate of 500 μN/s for Si and GaAs. **a** Young's modulus and **b** hardness. Te-Hua Fang, Win-Jin Chang and Chao-Ming Lin, Microelectronic Eng., 77, 389 (2005). With kind permission of Elsevier

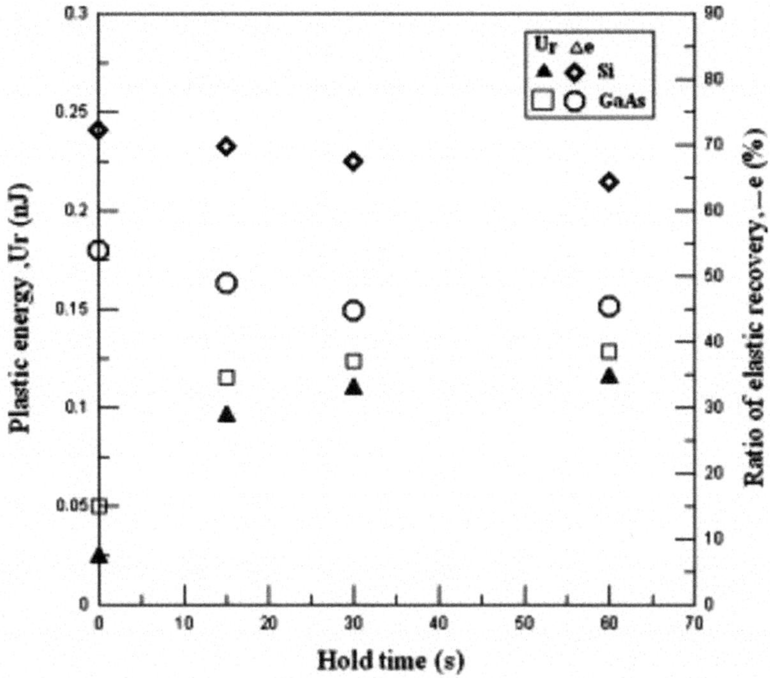

Fig. 12.23 The plastic energy and the ratio of elastic recovery versus the hold time at the load of 2000 μN and the loading rate of 500 μN/s. Te-Hua Fang, Win-Jin Chang and Chao-Ming Lin, Microelectronic Eng., 77, 389 (2005). With kind permission of Elsevier

12.5 Fatigue in Nanomaterials-Effect of Environment

12.5.1 Effect of Oxygen

12.5.1.1 In SiC

The presence of oxygen might induce oxidation in some material-in this case it is SiC, the subject of this section. During static and fatigue loading in SiC/SiC ceramic matrix ceramic (CMC) damage may occur in stress concentration region. It has been indicated throughout this book that SiC/SiC CMC is useful for various applications, among them aircraft engines. However, it has been observed that at high temperature use degradation of SiC/SiC components might occur. When hot airflow of ~2000 °C passes along the SiC/SiC CMC liner, a high gradient of temperature develops and generates high bending stresses. The SiC/SiC CMC liner has Y-shape discontinuity for fastening where fiber orientations change such that high stresses are concentrated locally and hence is more susceptible to damage during fatigue. In Fig. 12.24 a SiC/SiC CMC liner is shown schematically where the high stress-concentrated Y shape is seen. Recall that fatigue is a property which potentially determines material lifetime. Therefore, it is important to understand fatigue to avoid damage occurrence. Due to the inhomogeneous temperature distribution, the Y-shape region experiences (in addition to the high stress-concentration) cyclic bending stresses and which ultimately leads to fatigue damage. During the bending mode, the SiC/SiC CMC component fail at much lower fatigue cycles as compared to the conventional tensioncompression fatigue mode. However, at low fatigue stress levels where the components withstood 1000 fatigue run-out cycles, their degradation was still not severe. The typical transverse cross-section microstructures of 2.5D woven SiC/SiC CMC are shown in Fig. 12.25. 2a is the SEM microstructure showing various constituents of SiC/ SiC CMC. The composite is composed of SiC fibers of ~10 μm in diameter, which is coated with BN interphase of ~0.1 μm thickness along with the hybrid matrices fabricated by CVI and PIP processes (CVI is a chemical vapor infiltration process and PIP refers to polymer impregnation and pyrolysis technique). Shrinkage cracks were observed in the PIP matrix, which formed during the pyrolysis process, as shown in Fig. 12.25b. Some large pores in the PIP matrix as well as within the fiber tows were also observed.

To simulate the SiC/SiC CMC liner degradation at the high-stress-concentrated region inside the combustion chamber a unique point-load-bending test was used,

Annealing treatments were followed to simulate the actual application of several hours' flight. The tests were executed at 1000 °C at the heating rate of 12 °C/min and the displacement rate of 0.3 mm/min. For the experimental details the reader is referred to the original publication Patel et al. (2022) The fatigue run-out was set to be 1000 cycles.

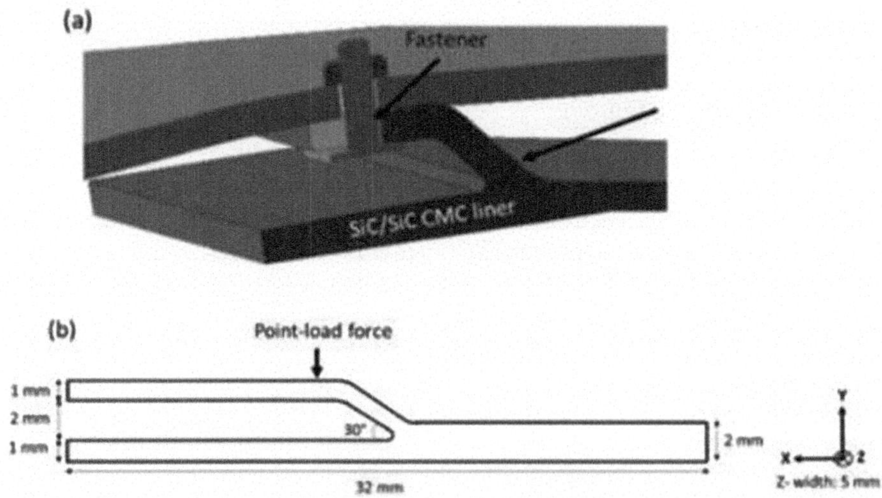

Fig. 12.24 **a** SiC/SiC CMC liner showing high-stress-concentrated Y-shape discontinuity and **b** the test sample dimensions. A. Patel, E. Sato, T. Takagi and N. Shichijo, J. Eur. Cer. Soc., 42, 4121 (2022). With kind permission of Elsevier

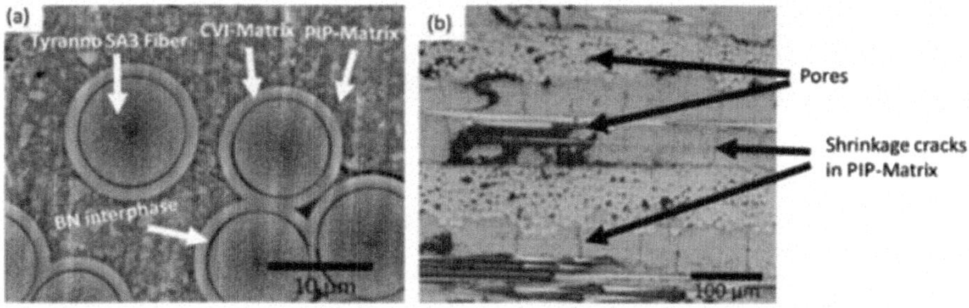

Fig. 12.25 The as-received SiC/SiC CMC component microstructures. **a** SEM microstructure showing various constituents of SiC/SiC CMC and **b** optical microscope image showing pores and shrinkage cracks in PIP-matrix. A. Patel, E. Sato, T. Takagi and N. Shichijo, J. Eur. Cer. Soc., 42, 4121 (2022). With kind permission of Elsevier

The largest stress, σ_L and the largest strain, ε_L at the Y-shape junction of the SiC/SiC CMC component were approximately calculated according

$$\sigma_L \approx 6Fl/bh^2 \quad (12.21)$$

$$\varepsilon_L \approx 6hy/4l^2 \quad (12.22)$$

where F is the applied load, l is the distance between the applied load and the Yahape junction, h is the sample thickness, b is the sample width and y is the crosshead displacement.

The fatigue stresses are plotted against failure cycles to obtain the SN curve in Fig. 12.26. As seen the fatigue stresses vs. the 1st matrix-cracking cycles (initial damage) are also plotted. Also note that the1 st matrix-cracking stress occurred at 230 MPa, lower than static strength. The list of the samples tested in bending fatigue is presented in Table 12.2. Typical bending-fatigue stress strain curves and the corresponding change in the normalized apparent stiffness are illustrated in Fig. 12.27. It can be seen that after the first cycle a large permanent strain is retained, which suggest that significant damage occurred in the early cycles (Fig. 12.27a). With increasing cycles, at the 54 th cycle, a deviation in the stress–strain curve confirmed the 1st matrix-cracking (Fig. 12.27b). Crack-opening and closing occurred with further cycles, and finally, just before the failure significant drop in the normalized apparent stiffness was observed (Fig. 12.27c). This drop in stiffness occurred when small matrix-cracking merged together to form a crack and then propagated rapidly, which led to the final failure. The microstructures of the 175 MPa-fatigue run-out sample after 1000 cycles at 1000 °C is shown in Fig. 12.28. The crack is initiated perpendicular to the fiber tows and propagates transversely along the fiber tows through the thickness of the specimen to a depth of 200 µm (Fig. 12.28c), side view microstructure). Figure 12.28b is a magnified top view of (a).

The presence of oxygen is manifested by oxidation, in particular at the interface where the BN constituent resides. Commonly BN is a preferential sintering additive.

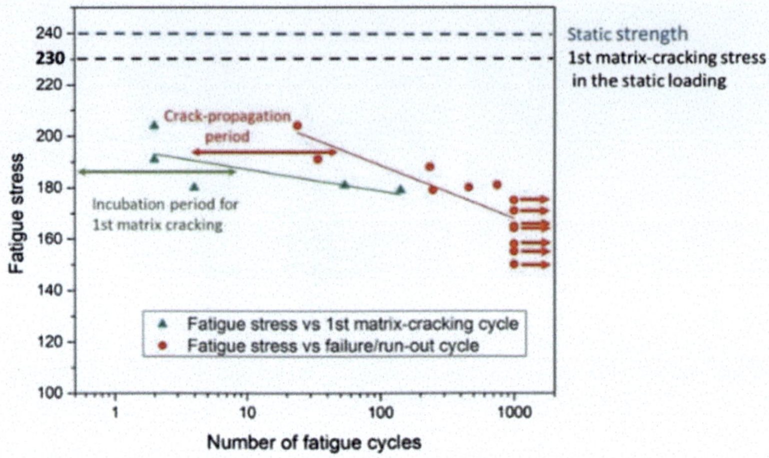

Fig. 12.26 Fatigue stress vs. 1st matrix-cracking cycle and failure/run-out cycle curves. A. Patel, E. Sato, T. Takagi and N. Shichijo, J. Eur. Cer. Soc., 42, 4121 (2022). With kind permission of Elsevier

Table 12.2 The list of samples and their corresponding fatigue properties executed in the bending-fatigue experiments. A. Patel, E. Sato, T. Takagi and N. Shichijo, J. Eur. Cer. Soc., 42, 4121 (2022). With kind permission of Elsevier

Sample number	Fatigue stress (MPa)	1st matrix cracking cycle	Failure fatigue 1000 (run-out) cycle	Retained post-fatigue strength (MPa)
1	204	**2**	24	–
2	191	**2**	34	–
3	188	–	235	–
4	181	54	748	–
3	180	4	461	–
6	179	142	247	–
7*	175	–	1000	218
6	171	–	1000	–
9*	165	–	1000	255
10	164	–	1000	–
11	158	–	1000	–
12	155	–	1000	–
13*	150	–	1000	282

Sample numbers 7*, 9*, and 13* were used to study the post-fatigue effect.

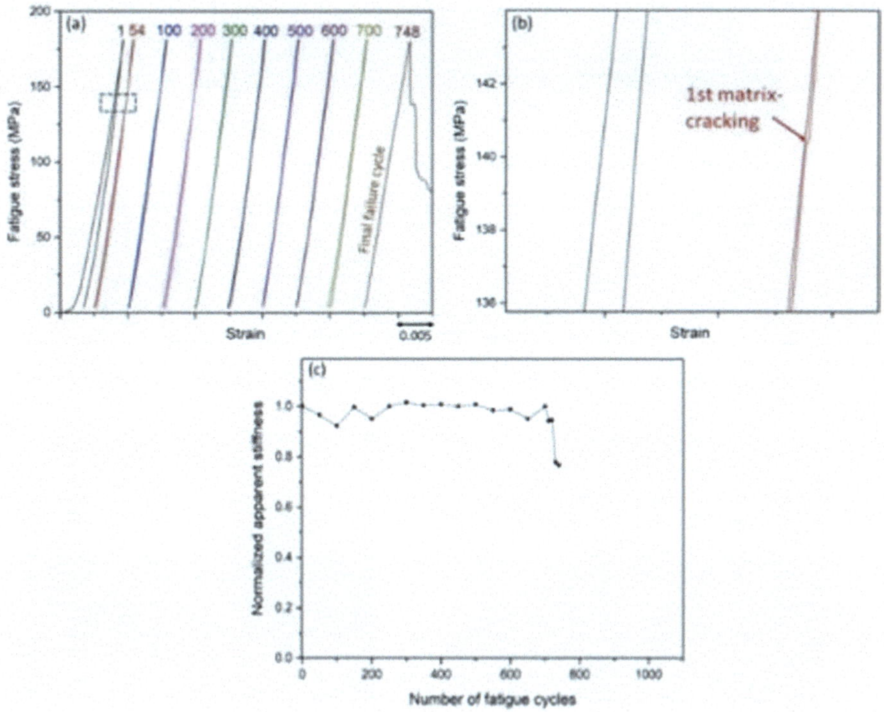

Fig. 12.27 **a** Typical bending-fatigue stress–strain curves (each curve was shifted to 0.005% for better visibility); **b** high magnification of rectangular dashed area showing 1st matrix-cracking and **c** the corresponding change in the normalized apparent stiffness against fatigue cycles. A. Patel, E. Sato, T. Takagi and N. Shichijo, J. Eur. Cer. Soc., 42, 4121 (2022). With kind permission of Elsevier

BN interphase degradation has a dual effect, namely: (i) formation of borosilicate glasses, which promote oxidation of SiC constituents, and (ii) filling of glassy phases in the pores within the fiber tows, which lead to the high retention of strength. To understand the effects of oxidation due to long-term fatigue at high-temperature on the mechanical properties of SiC/SiC CMCs is essential to ensure the best performance of SiC/SiC in actual use. It is already known that the matrix-cracking stress decreases with increasing test temperature and with increasing oxidation time. Generally, the fatigue lives are severely reduced for specimens tested in environments containing oxygen and/or water vapor (the effect of water vapor is more pronounced than the effect of oxygen).

In summary, fatigue in SiC/SiC CMC evaluated by point load bending was presented in this section. Fatigue tests were followed by annealing at 1000 °C for 50 h. Oxygen environment induces borosilicate formation leading to embrittlement and it also promotes SiO_2 formation resulting in some loss in the retained strength. The 1st matrix-cracking during fatigue tests is identified in the stress–strain curve as illustrated in Fig. 12.27.

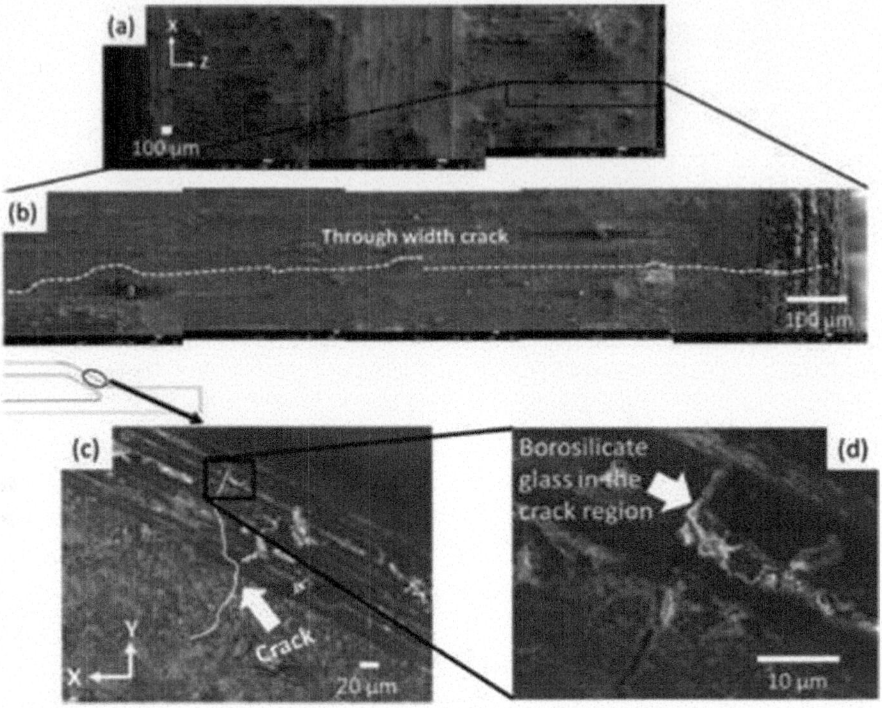

Fig. 12.28 Fatigue run-out microstructures showing top views **a**, **b** and side views **c**, **d** of the sample after fatigue at 175 MPa for 1000 cycles. A. Patel, E. Sato, T. Takagi and N. Shichijo, J. Eur. Cer. Soc., 42, 4121 (2022). With kind permission of Elsevier

12.5.2 Effect of Moisture

12.5.2.1 In Si

Contrary to bulk silicon, which is not susceptible to fatigue, micron-scale silicon does. This surprising behavior remains still unexplained completely. However, two major explanations were suggested Alsem et al. (2007). The first explanation attributes fatigue to a surface effect caused by subcritical cracking in the silicon-oxide layer that forms on the silicon sample and termed reaction-layer fatigue, the second suggests that subcritical cracking occurs in the silicon itself. One thing though is clear, that lower cyclic stresses result in larger number of cycles to failure in stress-lifetime data. However, existing experimental data seems to support the reaction-layer fatigue concept and provides explanation for the marked difference between the fatigue behavior of bulk and micron-scale silicon.

Fatigue behavior of single crystal and polycrystalline silicon in ambient air is shown in Figs. 12.29 and 12.30. In these figures a compilation of collected data are plotted as normalized stress versus number of cycles. A summary of fatigue testing of single-crystal and polycrystalline silicon thin films is listed in Tables 12.1 and 12.2 of Alsem et al. (2007) respectively. As mentioned, fatigue -by one explanation- is attributed to occur in the reaction-layer. The natural frequency of the resonator during cycling is prone to be damaged by environment. It was found that humidity has a critical effect on the damage accumulation rate and in 50% relative humidity (RH) damage magnitude is much larger than for example in 25% RH. Moreover, virtually no damage was observed in vacuum environment with a bake performed prior to the fatigue testing. Moisture-assisted subcritical cracking in a cyclic stress-assisted thickened surface oxide layer leads to failure in the reaction-layer by fatigue. The decrease in resonance frequency and the effect of moisture at various RHs, namely at 50, 25 and 40 RH is shown in Fig. 12.31. Notched specimens were used in these experiments. Usually, a 30 nm thick native oxide forms on the surface, which is uniformly distributed over the surface of the control specimen.

A native surface oxide of ~30 nm in thickness was uniformly distributed over the surfaces of the control samples. In contrast, the surface oxide layer was significantly thicker, up to ~90 nm at the notch root of fatigued specimens as seen in Fig. 12.32. It seems that the results indicate that fatigue of silicon thin films is attributed to a mechanism of sequential, cyclic stress-assisted oxidation and environmentally assisted cracking of the surface oxide layer which forms upon exposure to moisture and/

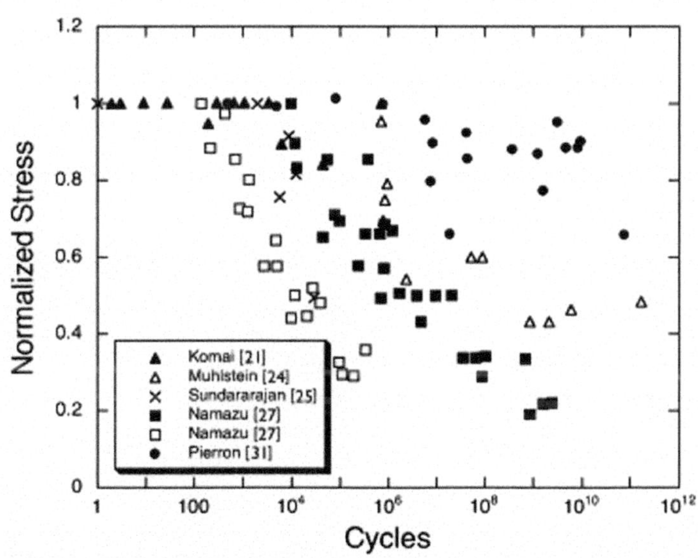

Fig. 12.29 Normalized applied maximum stress vs. number of cycles to failure (S–N) curve for the cyclic fatigue of single-crystal silicon in ambient air taken from the available literature. The stress values are normalized with respect to the stress from the test in that particular study that was run at the lowest number of cycles (and therefore in all but one case also with the test run at the highest stress). For the data from Namazu: [27] the open squares are tests in bending and the closed squares are tests in tension. D. H. Alsem, O. N. Pierron, E. A. Stach, C. L. Muhlstein and R. O. Ritchie, Adv. Eng. Mater., 9, 15 (2007). With kind permission of John Wiley and Sons

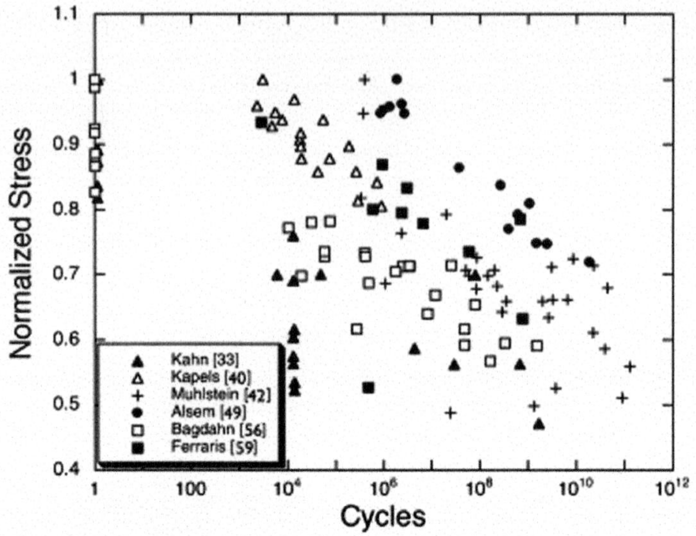

Fig. 12.30 Normalized, applied maximum stress vs. number of cycles to failure (S–N) data for fatigue of polysilicon in ambient air taken from the available literature. The stress values are normalized with respect to the stress from the test in that particular study that was run at the lowest number of cycles (and therefore in all but one case also with the test run at the highest stress). D. H. Alsem, O. N. Pierron, E. A. Stach, C. L. Muhlstein and R. O. Ritchie, Adv. Eng. Mater., 9, 15 (2007). With kind permission of John Wiley and Sons

or oxygen-containing atmospheres. This mechanism was termed above as reaction-layer fatigue. A schematic illustration of the reaction-layer fatigue mechanism for thin-film is illustrated in Fig. 12.33. Note the crack initiation in Fig. 12.33c. Damage accumulation is shown in Fig. 12.34. The maximum crack extension and final failure occurred in the oxide layer which can be seen in Fig. 12.33d, e. Since the crack in the oxide layer must cause failure of the entire structure, the criterion for this mechanism is that the thickness of the oxide layer, h, must be greater than or equal to the critical crack size, a_c, required to fail the entire structure, i.e., when $a_c < h$.

In summary, lifetimes of SiC/SiC CMC are found to be dependent markedly on the environment e.g., no fatigue failures are found in high vacuum, contrary to the effect of oxygen and/or moisture. It was found that the vast majority of experimental evidence on micron-scale, thin-film silicon fatigue is consistent with the concept of reaction-layer fatigue, where delayed failures result from cyclic stress-induced oxidation and consequent moisture-induced subcritical cracking in the silicon-oxide surface layer.

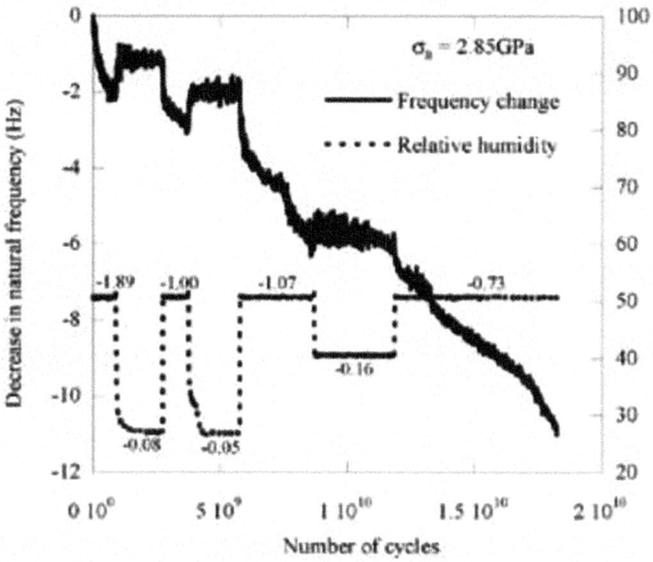

Fig. 12.31 Decrease in resonant frequency, f_0, during cycling at constant stress amplitude (2.85 GPa) in air (30 °C) at various successive relative humidity (RH) levels: 50, 25, 50, 25, 50, 40, and 50% RH. The numbers near the relative-humidity line indicate the average decrease in f_0 per 10^9 cycles for the particular humidity value. D. H. Alsem, O. N. Pierron, E. A. Stach, C. L. Muhlstein and R. O. Ritchie, Adv. Eng. Mater., 9, 15 (2007). With kind permission of John Wiley and Sons

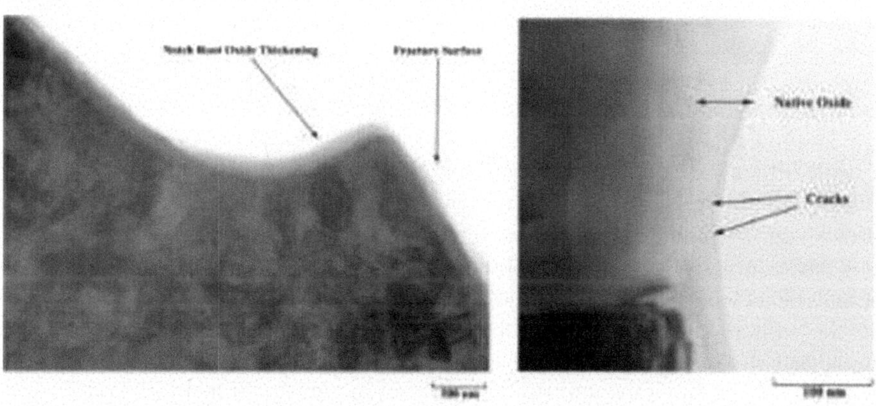

Fig. 12.32 HVTEM images of the notch region in an untinned, 2 mm thick, polycrystalline silicon test sample after high-cycle fatigue. (Left) This image shows enhanced oxidation at the notch root that failed, after 3.56×10^9 cycles at stress amplitude of $\sigma_a = 2.26$ GPa. (Right) This image shows stable cracks, ~50 nm in length, in the native oxide formed during cyclic fatigue loading; testing of this sample was interrupted after 3.56×10^9 cycles at a stress amplitude $\sigma_a = 2.51$ GPa. Image was intentionally defocused to facilitate the observation of the crack. D. H. Alsem, O. N. Pierron, E. A. Stach, C. L. Muhlstein and R. O. Ritchie, Adv. Eng. Mater., 9, 15 (2007). With kind permission of John Wiley and Sons

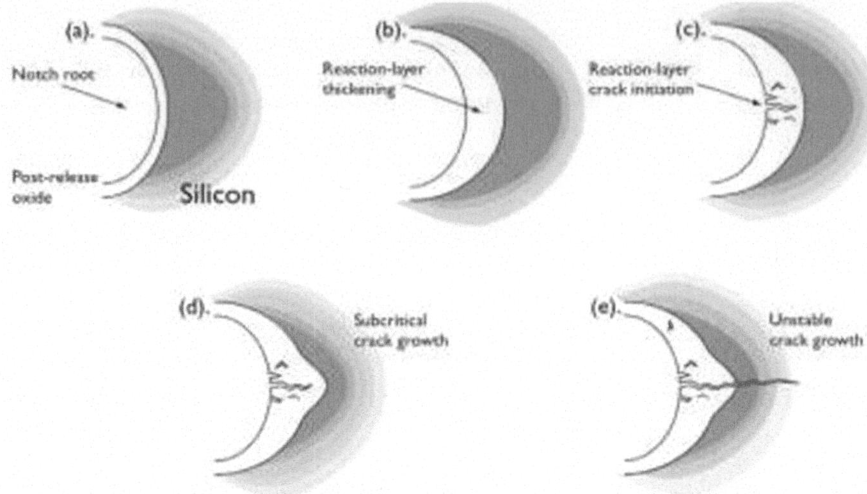

Fig. 12.33 Schematic illustration of the reaction-layer fatigue mechanism for thin-film fatigue at the notch of the polycrystalline silicon cantilever beam: **a** reaction layer (post-release oxide) on surface of the silicon, **b** localized cyclic stress-assisted oxide thickening at the notch root, **c** moisture-assisted crack initiation in the surface oxide at the notch root, **d** additional thickening and cracking of reaction-layer, and **e** unstable crack growth in the silicon film. D. H. Alsem, O. N. Pierron, E. A. Stach, C. L. Muhlstein and R. O. Ritchie, Adv. Eng. Mater., 9, 15 (2007). With kind permission of John Wiley and Sons

12.5.2.2 In SiC

The lack of material, and the importance of the topic in many industries, in particularly in aviation, calls for adding the macro-size subject on fatigue in SiC despite the possible difference in mechanical behavior between nano and bulk material.

Rather than SiC, a SiC matrix composite (CMC), SiC/SiC is considered in this section with a BN interface addition. The effect of moisture on fatigue behavior of this CMC composite at elevated temperature is of special interest due to the demand for advanced materials in aerospace and other high-temperature application. The SiC/SiC system used for fatigue resistance evaluation is a NASA developed type termed Syl-iBN/BN/SiC where Syl (stands for Sylramic) is a SiC fiber. This CMC system is one of strongest and the most creep resistant systems. For experimental details on the production of Syl-iBN/BN/SiC consult Shankar (2005).

Fatigue tension-tension tests of the composite are compared with monotonic tensile tests. Monotonic stress -strain relations at RT and 750 °C for 0% and 60% moisture are illustrated in Fig. 12.35. It is seen that the strength is dependent on the temperature, but no dependence on moisture content could be observed at 750 °C. Overall, the stress–strain relationships are similar at all test conditions except

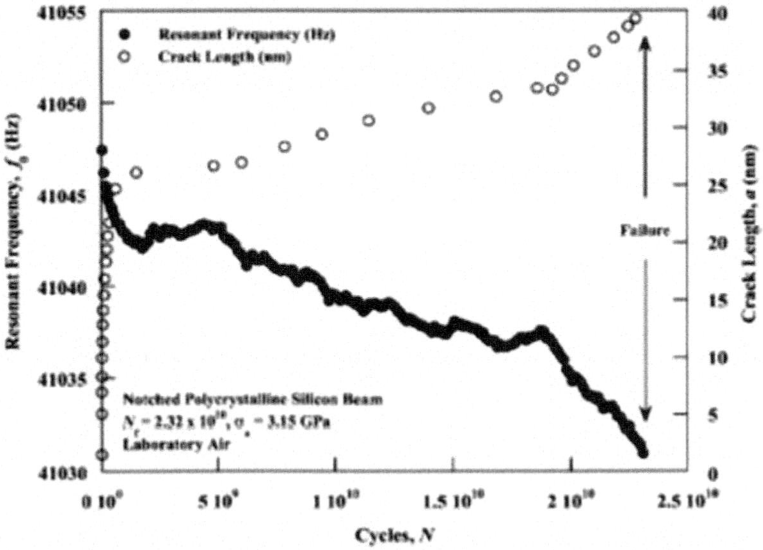

Fig. 12.34 Representative damage accumulation in polycrystalline silicon, shown by experimentally measured decrease in resonant frequency, f_{crack}, with cycles during a fatigue test ($N_f = 2.23 \times 10^{10}$ cycles at $\sigma_a = 3.15$ GPa and the corresponding computed increase in crack length, a. D. H. Alsem, O. N. Pierron, E. A. Stach, C. L. Muhlstein and R. O. Ritchie, Adv. Eng. Mater., 9, 15 (2007). With kind permission of John Wiley and Sons

for a variation of about 20% in the ultimate tensile strength. The stress–strain relationship shows initially a linear portion up to about 200–250 MPa, which represents the part before any appreciable matrix crack, develops in the composite. Compare the monotonic strength with the normalized fatigue life diagram shown in Fig. 12.36. The normalization is of the maximum (peak) applied cyclic stress to the average composite's monotonic strength of 345 MPa at 750 °C. The figure shows the detrimental effect of moisture (RH-60%) relative to dry environment (RH-0%).

Strain variation in dry and humid environment is shown for 0% and 60% RHs, respectively in Fig. 12.37. In Fig. 12.38, the Young's moduli near the failure point of the fatigue tests for the test conditions applied are compared. It is seen from this figure that more damage is present in humid environment (RH 60%) than in dry with.

0% moisture. The specimen had less stiffness at a given cyclic stress level at or near failure under the humid test condition than that under dry test condition.

Fractured specimens evaluated under monotonic tension and fatigue tests under both humid and dry test environments were evaluated with optical and scanning electron microscopes. These are compared in a series of illustrations in Figs. 12.39, 12.40 and 12.43. In Fig. 12.39 an overall pattern of failure in the gage region at low magnification is shown. No distinctive feature can be attributed to the environmental condition (i.e., whether dry or humid) at low magnification. However, at monotonic.

tension tests at the higher magnification there was neither a noticeable difference on the fracture surface between the dry and humid conditions. Under wet

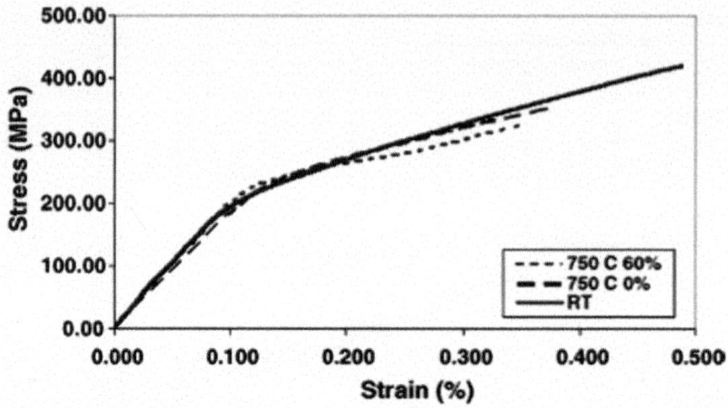

Fig. 12.35 Monotonic stress–strain relationships. M. Shankar, Mater. Sci. Eng. A 412, 165 (2005). With kind permission of Elsevier

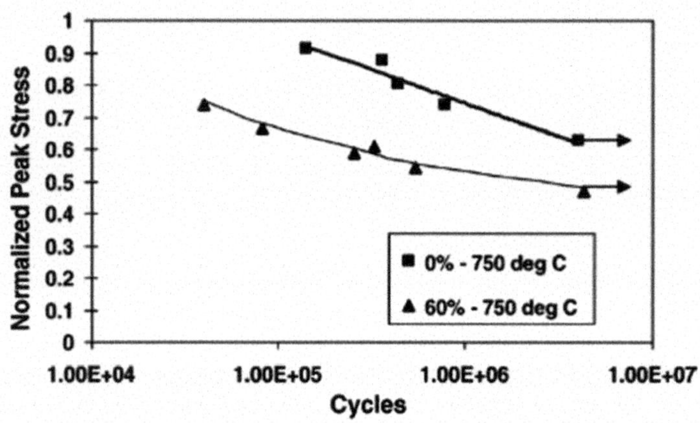

Fig. 12.36 Applied stress vs. cycles to failure relationships. M. Shankar, Mater. Sci. Eng. A 412, 165 (2005). With kind permission of Elsevier

and dry conditions, the fracture surfaces were rough and flat indicating brittleness. Fiber pullout is seen -a pattern familiar from brittle CMC-. A comparison of fracture surfaces from fatigue tests under dry and humid test conditions are shown in Figs. 12.41 and 12.42, respectively. Under dry test condition the fracture surface was relatively rougher, and more fiber pull-out was observed. Further, Figs. 12.43 and 12.44 clearly show embrittlement at higher magnification at the fiber/matrix interphase in dry condition and in the presence of moisture, respectively. The BN interphase between fiber and matrix showed evidence of brittle fracture with no evidence of degradation in the dry test condition (Fig. 12.43). It appears however, that in the humid environment the BN interphases were oxidized (Fig. 12.44). The

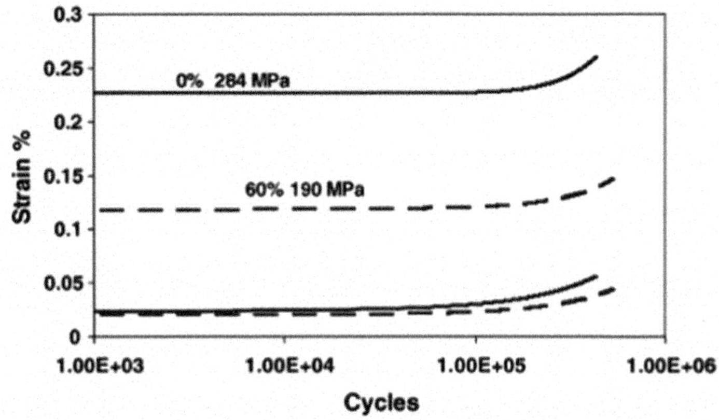

Fig. 12.37 Typical evolution of strain during fatigue test (upper and lower curves represent maximum and minimum strains). M. Shankar, Mater. Sci. Eng. A 412, 165 (2005). With kind permission of Elsevier

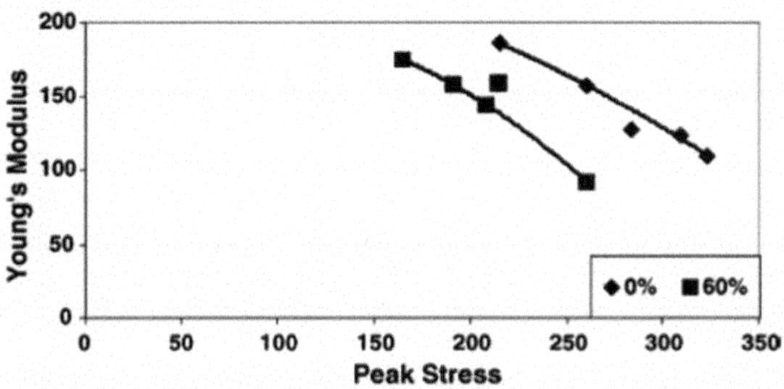

Fig. 12.38 Young's modulus vs. applied stress at the end of fatigue test. M. Shankar, Mater. Sci. Eng. A 412, 165 (2005). With kind permission of Elsevier

microscopic analysis suggests that embrittlement occurred in the composite which causes reduction in fatigue in the presence of moisture relative to the test condition.

In summary, the results of the SiC/SiC composite as the Syl-iBN/BN/SiC system is presented. This CMC was investigated at an elevated temperature in the presence of moisture and also compared with the system tested at dry environment. Fatigue life at a prescribed stress level was shorter in the presence of moisture than that without moisture (0% RH). Microscopic analysis indicates that the presence of moisture in the test environment causes embrittlement of the BN interphase between fiber/matrix.

or/and fiber/fiber locations. It has been observed that moisture caused embrittlement of the interphase between fiber and matrix, which in turn caused the degradation of fatigue performance of the SiC/SiC CNC.

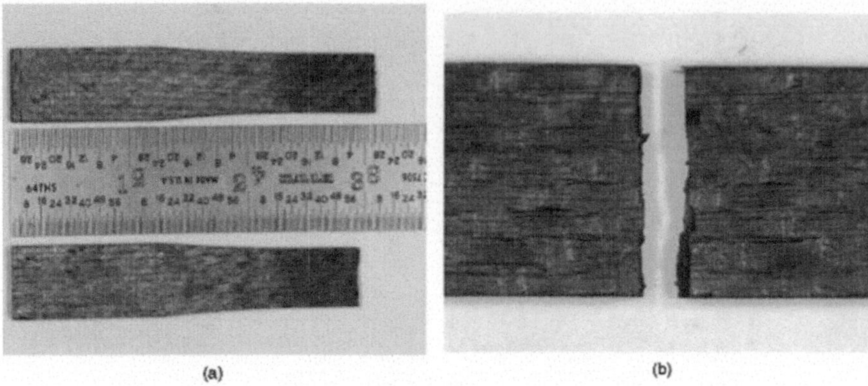

Fig. 12.39 Typical fractured specimen; **a** overall view and **b** close-up of fracture. M. Shankar, Mater. Sci. Eng. A 412, 165 (2005). With kind permission of Elsevier

Fig. 12.40 Close-up of fracture surface (monotonic test). M. Shankar, Mater. Sci. Eng. A 412, 165 (2005). With kind permission of Elsevier

12.6 Fracture in Nanomaterials-Effect of Environment

12.6.1 Effect of Oxygen

12.6.1.1 In Si

At room temperature, both single crystal silicon (SCS) and polysilicon are brittle, and their strength is associated with the applied stress required to propagate a crack across a component with the consequent reduction of their load-carrying capacity. In

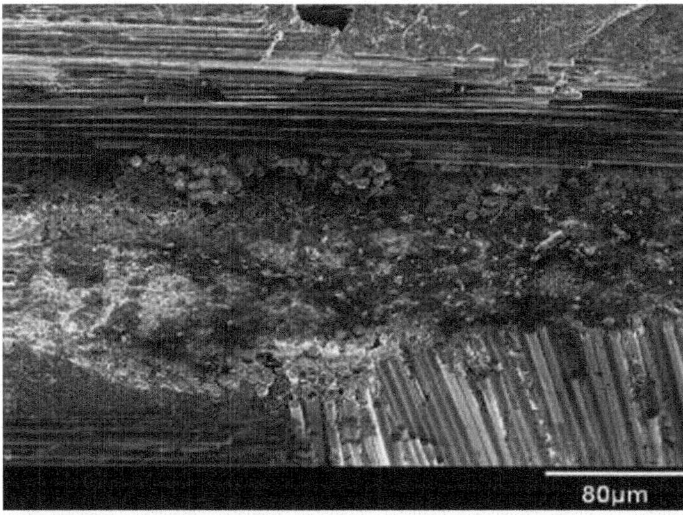

Fig. 12.41 Close-up of fracture surface (fatigue test under dry condition). M. Shankar, Mater. Sci. Eng. A 412, 165 (2005). With kind permission of Elsevier

Fig. 12.42 Close-up of fracture surface (fatigue test under humid condition). M. Shankar, Mater. Sci. Eng. A 412, 165 (2005). With kind permission of Elsevier

general, the strength depends on the resistance of the component to fracture which is related to the flaws from which cracks can grow.

Silicon nanowires (NWs) are one of the most important nanostructures used for fabricating various electronic and optoelectronic nanodevices. Important issue is their mechanical properties such as the elastic, plastic, and fracture properties in their applications as nanostructures. Only limited information is available about the structure–mechanical property correlations of Si. The surface of a Si NWs may be

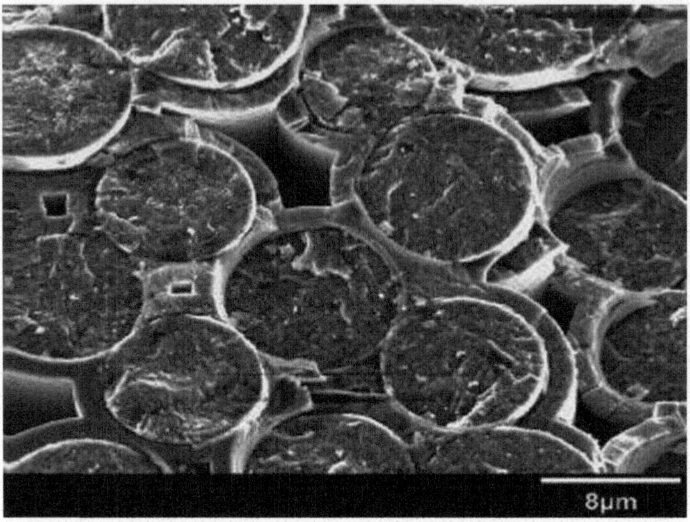

Fig. 12.43 Close-up of fiber/matrix interphase from fatigue test under dry condition; maximum stress = 284 MPa, cycles to failure = 4.35E06, 750 °C. M. Shankar, Mater. Sci. Eng. A 412, 165 (2005). With kind permission of Elsevier

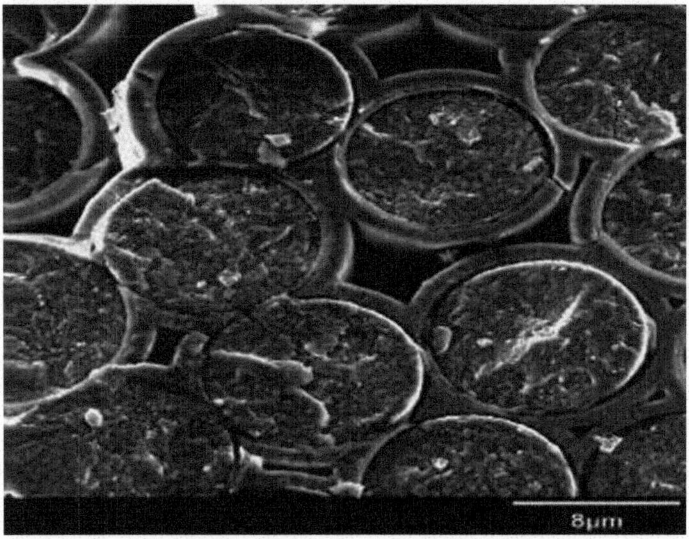

Fig. 12.44 Close-up of fiber/matrix interphase from fatigue test under humid condition; maximum stress = 165 MPa, cycles to failure = 4.25E06, 750 °C. M. Shankar, Mater. Sci. Eng. A 412, 165 (2005). With kind permission of Elsevier

covered by a thin oxide layer, and it is important to investigate the role possibly played by oxygen diffusion into the lattice, especially in the strained area, during plastic deformation. In the experiment conducted, in situ electron energy loss spectroscopy (EELS) in an attachment to the transmission electron microscope was used; data were acquired simultaneously while conducting the axial tensile tests on the Si NW. EDS line scans of silicon and elemental oxygen across the Si NW prior to and after necking deformation is illustrated in Fig. 12.45. In Figs. (a) and (b) two sets of EELS spectra taken from a Si NW are shown before and after superplasticm deformation. Figure 12.45a shows a series of EELS spectra taken from the Si NW prior to plastic deformation. Figure 12.45b shows the corresponding EELS spectra taken from the superplastically deformed Si NW with a neck. In Figs. 12.45c, d the locations at which the EELS spectra were taken and from comparison of the Si ledge with standard samples, a 3.5 nm thick SiO_x surface layer was revealed (identified by EELS spectra of 12,45 (a). The SiO_x has amorphous feature. EDS line scans of the Si NW prior to and after deformation. Figure 12.45 g, h show the K-edge spectra from the in situ EDS line scans of silicon and elemental oxygen prior to and after plastic deformation, namely prior to and after necking.

From axial tensile experiments, plasticity at ambient temperature can be correlated with LSP of the dislocation velocity and the Young's modulus according to

$$v = A\tau^m \exp\left[\frac{-U}{K_B T}\right] \tag{12.23}$$

where v is dislocation velocity, A and m are constants, t is shear stress, k_B is the Boltzmann constant, T is absolute temperature, and U is the activation energy of a dislocation. The activation energy can be regarded as a constant that reflects the atomic–atomic interactions and bonding characters and can be correlated to the shear and elastic moduli G and E, respectively by

$$\mathrm{U} \infty G a_0 \tag{12.24}$$

$$U \infty E a_0 / [2(1+\gamma)] \tag{12.25}$$

where a_0 is the lattice constant and γ is the Poisson ratio. The NW's surface structure strongly affects the NW's physical, chemical, and mechanical properties. The elastic modulus of Si is related to the size of the NW, and it decreases with decreasing NW diameter and film thickness (Fig. 12.46). As seen in the figure also, the elastic modulus of Si is related to the size of the NW, and it decreases with decreasing NW diameter and film thickness. Assuming that a_0 and c are constants, we can write a relation

$$U_N = \frac{E_N}{E_0} U_B \tag{12.26}$$

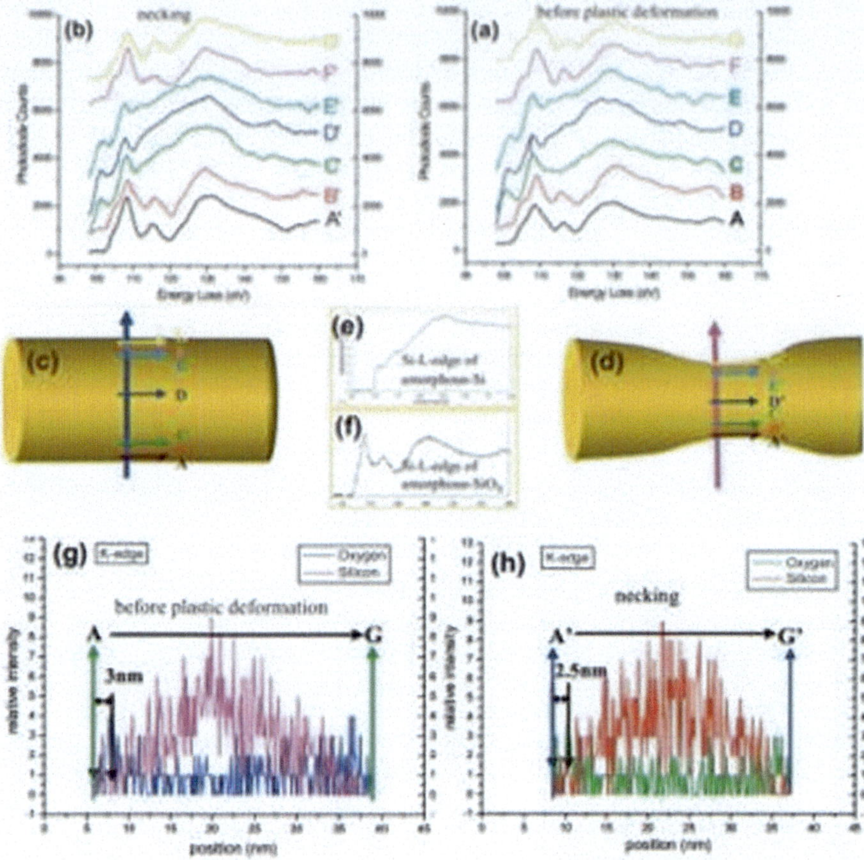

Fig. 12.45 Series of EELS spectra and EDS line scans taken from a Si NW prior to and after large-strain plastic deformation. Both series of EELS spectra were taken across the NW width. **a** A series of spectra taken from the Si NW prior to plastic deformation. **b** A series corresponding to the NW with a superplastic necking feature. **c, d** Schemes of the locations at which the spectra were taken for (a) and (b), respectively. (e, f) Schulmeister et al. (2003) EELS spectra for the Si L edge of Si and amorphous SiO_x, respectively. (g, h) Line scan EDS spectra of elemental oxygen and silicon for the Si NW before and after neck formation. The surface oxygen layer thickness was measured as the distance from the starting position A or A¢ to the first peak position of oxygen. X. Han, K. Zheng, Y. F. Zhang, X. Zhang, Z. Zhang and Z. L. Wan, Adv. Mater., 19, 2112 (2007). With kind permission of John Wiley and Sons

where UB and EB are the dislocation activation energy and Young's modulus of bulk silicon, respectively, and U_N and E_N are those of the NW. Then we obtain the dislocation velocity in the NW as

$$v_N = A\tau_N^m \exp\left[\frac{-E_N U_B}{K_B K_B T}\right] \quad (12.27)$$

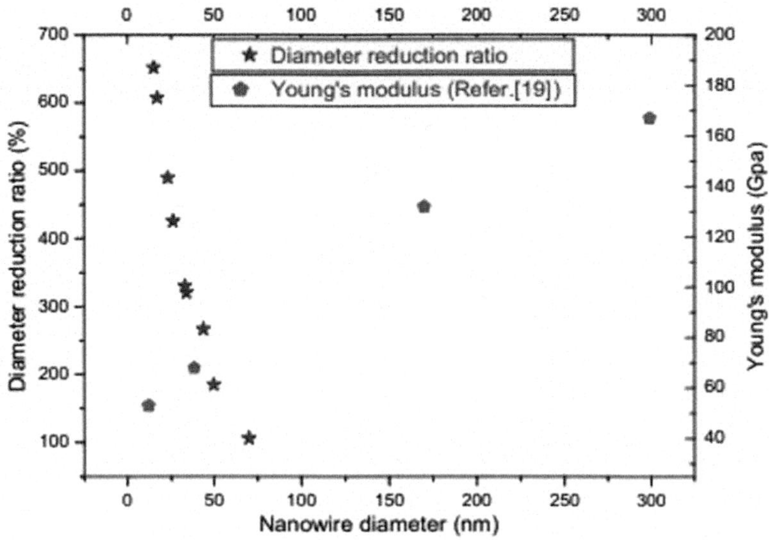

Fig. 12.46 The diameter reduction ratio as a function of nanowire diameter in the case of axial tension at a low temperature close to room temperature. Young's modulus values as a function of diameter are plotted in the same figure for comparison. X. Han, K. Zheng, Y. F. Zhang, X. Zhang, Z. Zhang and Z. L. Wan, Adv. Mater., 19, 2112 (2007). With kind permission of John Wiley and Sons

Using $U_B = 2.2$ eV and $E_B = 169$ GPa, we obtain the correlation of dislocation activation energy for the NWs as shown in Fig. 12.12 and 12.47a. It is seen that the dislocation activation energy, U_N with decreasing NW diameter, being lower (0.72 eV) than that of the bulk material (2.2 eV). In Fig. 12.47b the normalized dislocation velocity ratio v/v_{av} is plotted as a function of the Si NW diameter, where v_{av} is the average dislocation velocity for Si NWs.

In summary, room-temperature large strain plasticity (LSP) of single-crystalline Si.

NWs was directly observed by axial tensile experiments carried out in situ in an ultrahigh-resolution electron microscope. A very large reduction in NW diameter occurred before fracture which is about three orders of magnitude higher than that of bulk Si. The ambient temperature Si NW tensile plasticity was initiated by the emergence of a high density of dislocations, followed by the development of a continuous disordered lattice by the emission of dislocations and formation of disordered crystalline structures, before it finally became amorphous and the plastic formation of long necks occurred, which eventually resulted in fracture. The sizedependent fracture mechanism is attributed to the scale-related dislocation activities. The plasticity of a particular material (i.e., whether it is brittle or ductile) is controlled by the dislocation velocity, which is regarded as the rate of emission of dislocations and/or dislocation motion.

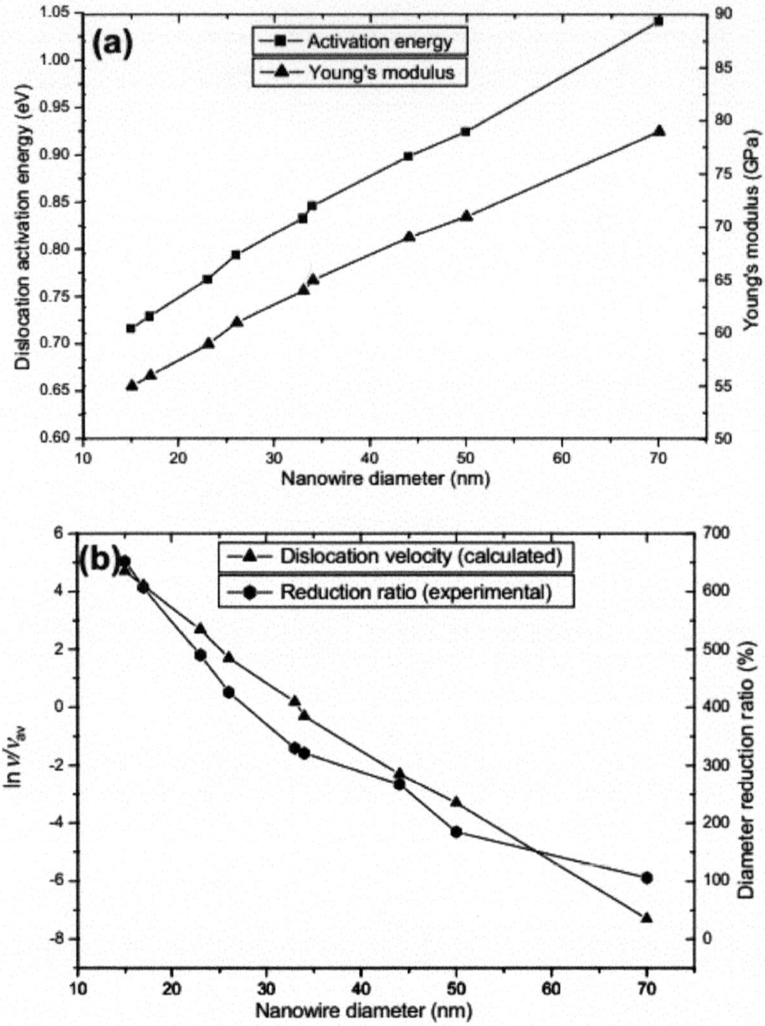

Fig. 12.47 **a** The dislocation activation energy of nanowires U_N with respect to the diameter of the nanowires. The extrapolated Si NW Young's modulus is also plotted. **b** The normalized dislocation velocity ratio v/v_{av} as a function of the diameter of the Si nanowires. The diameter reduction ratio is plotted in the same figure for comparison. X. Han, K. Zheng, Y. F. Zhang, X. Zhang, Z. Zhang and Z. L. Wan, Adv. Mater., 19, 2112 (2007). With kind permission of John Wiley and Sons

12.6.1.2 In SiC

For the developing of high quality SiC nanoceramics, the detrimental effect of oxygen has to be limited. Oxygen effects the SiC product quality by coarsening and limiting densification. Sintering in the 1700–1900 °C range produces different microstructures. Reduction of SiO_2 during sintering was observed at 1500 °C according to the following reactions leading to SiO formation

$$SiO_2 + C\ (s) \rightarrow CO\ (g) + SiO\ (g) \quad (12.28)$$

$$2SiO_2 + SiC\ (s) \rightarrow CO\ (g) + 3SiO\ (s) \quad (12.29)$$

The result of the sintering is shown in Fig. 12.48 for sample prepared by two methods, namely, magnetic stirring (MS) and ball milling (BM). The microstructures of the as produced powder, and the fractured surfaces of the magnetically stirred and ball milled samples are shown in Figs. 12.49, 12.50 and 12.51.

The physico-chemical properties of the as-produced powder (Fig. 12.49) are listed in Table 12.3, and the oxygen content of the sintered samples is shown in Table 12.4.

No significant difference exists in the relative densities between the magnetic stirring and ball milling samples as illustrated in Fig. 12.52.

For the correlation of microstructure and mechanical behavior, some mechanical characterization of the samples, such as hardness and toughness were performed, and which are shown in Figs. 12.53 and 12.54. The influence of grain size affects hardness through the well-known Hall–Petch relationship given as

$$H_V = \sigma_0 + Kod^{-1/2} \quad (12.30)$$

where H_v is the Vickers hardness, σ_0 and k_0 are constants and d is the grain size, respectively. The grain size can be estimated from the specific surface area (SSA)

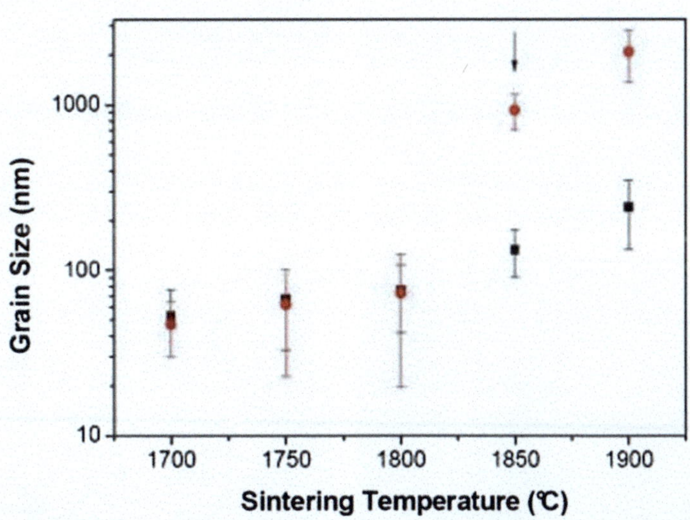

Fig. 12.48 Evolution of the grain size as a function of the sintering temperature (■, magnetic stirring; •, ball milling). The arrow indicates the excessive grain growth. F. Lomello, G. Bonnefont, Y. Leconte, N. Herlin-Boime and G. Fantozzi, J. Eur. Ceram. Soc., 32, 633 (2012). Open access

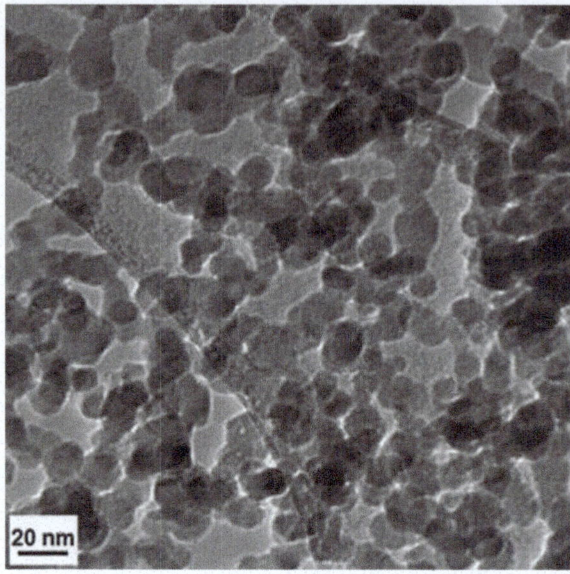

Fig. 12.49 TEM micrograph of the as-produced powder. F. Lomello, G. Bonnefont, Y. Leconte, N. Herlin-Boime and G. Fantozzi, J. Eur. Ceram. Soc., 32, 633 (2012). Open access

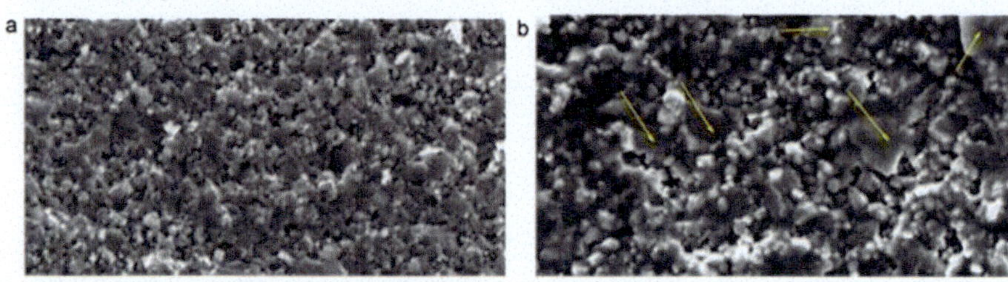

Fig. 12.50 SEM micrographs at the same magnification (100 k ×) of the fracture surfaces of magnetically-stirred samples (MS): Sintered at 1750 °C (a) and 1900 °C. F. Lomello, G. Bonnefont, Y. Leconte, N. Herlin-Boime and G. Fantozzi, J. Eur. Ceram. Soc., 32, 633 (2012). Open access

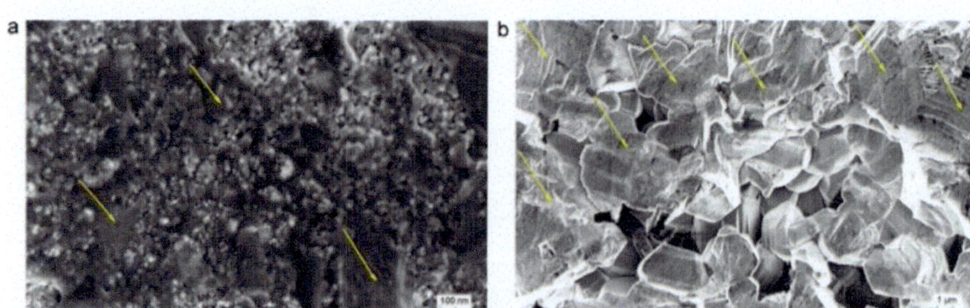

Fig. 12.51 SEM micrographs of the fracture surfaces at different magnifications (100 k × and 10 k ×) of ball-milled samples (BM): sintered at 1750 °C (a) and 1900 °C (b). F. Lomello, G. Bonnefont, Y. Leconte, N. Herlin-Boime and G. Fantozzi, J. Eur. Ceram. Soc., 32, 633 (2012). Open access

Table 12.3 Physico-chemical properties of the as-produced powder. F. Lomello, G. Bonnefont, Y. Leconte, N. Herlin-Boime and G. Fantozzi, J. Eur. Ceram. Soc., 32, 633 (2012). Open access

Average particle size (TEM)	16.6 nm
SSA(BET)	114m^2/g
True density	3.02 g/cm^3
Crystal phase	β
Si (wt%)	65.7%
C (wt%)	29.2%
O (wt%)	5.0%

Table 12.4 Oxygen content of dispersed and sintered samples. F. Lomello, G. Bonnefont, Y. Leconte, N. Herlin-Boime and G. Fantozzi, J. Eur. Ceram. Soc., 32, 633 (2012). Open access

Sample	Oxygen content (wt%)
Dispersed MS	6.1
MS sintered at 1900 C	0.7
Dispersed BM	6.6
BM sintered at 1900 C	0.8

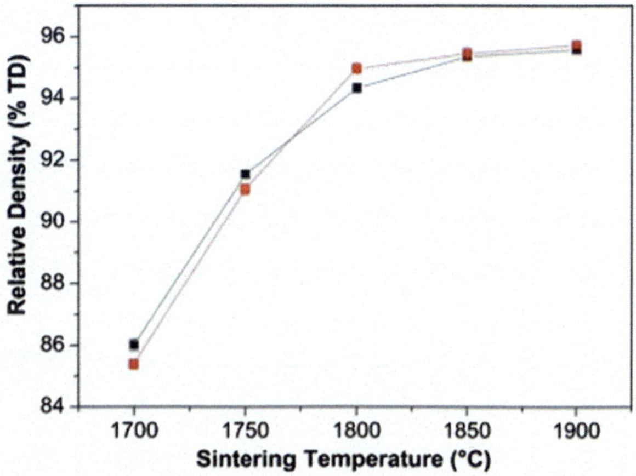

Fig. 12.52 Evolution of the density as a function of the sintering temperature (■, magnetic stirring; •, ball milling). F. Lomello, G. Bonnefont, Y. Leconte, N. Herlin-Boime and G. Fantozzi, J. Eur. Ceram. Soc., 32, 633 (2012). Open access

measured as 114 m^2/g. Assuming a spherical shape and using the true density ρ, a corresponding grain size D can be estimated by

$$D = \frac{6}{\rho ssA} \tag{12.31}$$

The fracture toughness. K_{Ic} is an important design parameter and is determined by

$$K_{Ic} = A\left(\frac{E}{H}\right)^{\frac{1}{2}}\left(\frac{p}{C^{\frac{2}{3}}}\right) \tag{12.32}$$

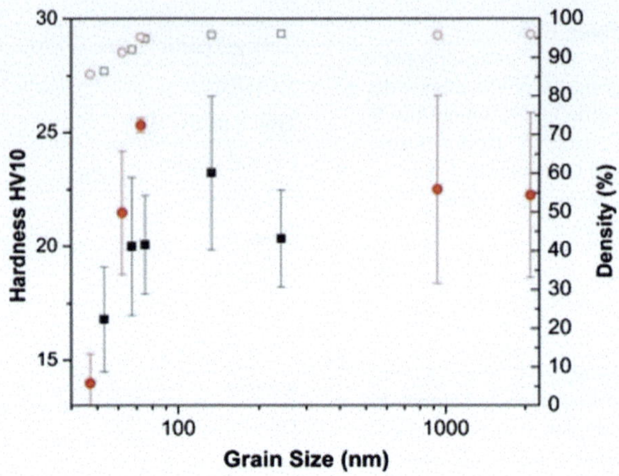

Fig. 12.53 Variation of Vickers hardness and relative density as a function of grain size (■ = magnetic stirring, • = ball-milling – full colored: hardness, noncolored: density). F. Lomello, G. Bonnefont, Y. Leconte, N. Herlin-Boime and G. Fantozzi, J. Eur. Ceram. Soc., 32, 633 (2012). Open access

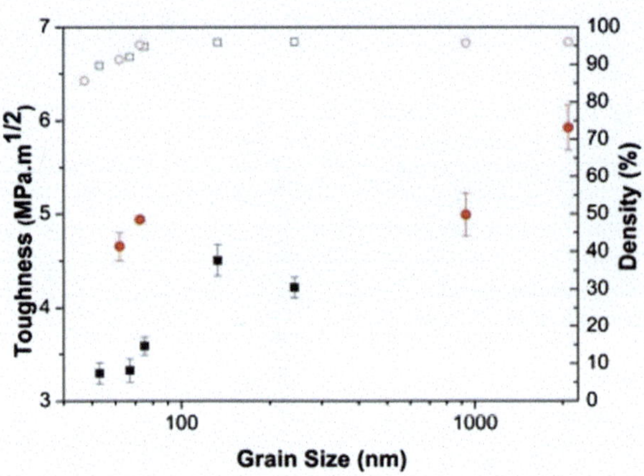

Fig. 12.54 Variation of fracture toughness and relative density as a function of grain size (■ = magnetic stirring, • = ball-milling – full colored: hardness, noncolored: density). F. Lomello, G. Bonnefont, Y. Leconte, N. Herlin-Boime and G. Fantozzi, J. Eur. Ceram. Soc., 32, 633 (2012). Open access

where c is the length of the crack from the center of the impression, A is a geometric constant equal to 0.016, P is the change expressed in Newtons, H is the hardness value and E the Young's modulus. For the calculation of the elastic modulus the variation as a function of porosity, Vp has been considered and given as

$$E = E_0 \exp(-CV_p) \quad (12.33)$$

where E_0 is the elastic modulus of the pore free SiC given as 460 GPa, and C is a constant of 3.57. Samples with grains over 200 nm exhibited the highest toughness of 6 ± 0.2 MPa.m$^{1/2}$ as seen in Fig. 12.54.

In summary, good densification (~96%) was obtained in the consolidation of SiC nano powders by laser pyrolysis and sintered by spark plasma sintering. This technique is performed without sintering additives. It was observed that grain size and porosity affect the mechanical properties and to enhance these properties it is essential to control the oxygen content in the nano carbide ceramics. By preparing SiC powders in controlled atmosphere surface oxidation can be avoided.

12.6.2 *Effect of Moisture*

12.6.2.1 In Si

Devices based on single crystal silicon (SCS) are of interest in microelectronics and microelectromechanical systems (MEMS) for device application and their fabrication is a critical step to create surfaces free of stress concentrations as much as possible. At such sites failure initiation may occur leading to premature fracture. Silicon microfabrication techniques have enabled the development and commercialization of (MEMS).

Humidity has also been shown to have a marked effect on reducing the resistance to fracture in particular that of fatigue, which was observed in polysilicon also DelRio et al. (2015).This degrading effect of humidity has been attributed to a reaction layer process involving surface oxide thickening. In air the oxide layer is less tough than polysilicon. Further, The surface oxide, when under residual stress, is known to be susceptible to stress corrosion cracking under ambient (humid) environments. As a result, thin film tension tests of microfabricated SiO_2 show that both the strength and toughness is substantially less in humid air than in vacuum. In addition, significant stresses develop caused by large volumetric mismatch between silicon and its oxide. The large stresses resulting from the volumetric mismatch are lessened, but not completely eliminated by the formation of defects including dislocations, atomic ledges, and nanoscale cracks. There is a connection between fracture energy and toughness for SCS and polysilicon which is given by DelRio et al. (2015)

$$\mathrm{T}_{(\mathrm{hkl})[\mathrm{uvw}]} = (2_{\gamma(hkl)}\mathrm{M}_{(hkl)[uvw]})^{1/2} \tag{12.34}$$

where T is the toughness (often designated Kc, and for isotropic materials taken to be a material constant) and M(hkl)[uvw] is an elastic fracture factor. The predicted toughness values for SCS for the three major fracture planes are $0.87 < \mathrm{T}(001) < 0.89$ MPa $^{\mathrm{m}1/2}$, $0.73 < \mathrm{T}(110) < 0.80$ MPa $\mathrm{m}^{1/2}$, and $\mathrm{T}(111) \approx 0.72$ MPa $\mathrm{m}^{1/2}$ based on the parameters indicated in the publication DelRio et al. (2015). An almost invariant toughness as a function of crack orientation is evaluated. It is believed that moisture plays a strong role in the crack initiation and propagation process which affects the fracture strength. Regarding fatigue, lifetimes were significantly reduced for specimens tested in environments with water vapor and oxygen, with the effect of water vapor much more pronounced than that of oxygen. Further, the resonant frequency remained constant in dry air, but decreased dramatically in wet air (moisture) which played a strong role in the crack initiation and propagation process. For detailed experimental analysis and methods used one is referred to DelRio et al. (2015).

In summary, humidity has a marked effect on reducing the resistance to fracture in particular that of fatigue, which was observed in polysilicon also. The resonant frequency decreases significantly in moisture. The increased use of MEMS based on SCS is related to the improvement of the mentioned shortcomings.

12.6.2.2 In SiC

SiC/BN/SiC CMC degradation under low temperature water exposure is presented in this section on the environmental effect on the mechanical properties, specifically on fracture. Humid environments can be detrimental for BN interphases, leading to a drop in mechanical properties. Furthermore, initial low-temperature humid environments can induce a faster degradation during subsequent thermal exposure. BN is a common interface (or pyrolytic carbon also) and is expected to ensure load transmission to the fibers and promoting crack deflection resulting in a tougher material. For aerospace applications, BN is the best choice due to its ability to form liquid oxide (B_2O_3) instead of volatile species (CO or CO_2) under high temperature in oxygen-containing environments.

The fiber used is Nicalon coated with CVD boron nitride and 5-Harness Satin Weave (5HS) fabric also coated with BN. The samples used for the mechanical testing were cut into dog bone specimens and exposed to different degradation environments. The following degradation steps have been used simulating an aeroengine environment:

1. Exposing in air to 800 °C for 500 h at a relative humidity of 31% at 23 °C. Heating and cooling rates 10 °C min^{-1}.
2. Water exposure for 500 h at 65 °C in a moist environment with 95% relative humidity (corresponding to partial pressure of water of 0.23 atm).
3. Reheating the water exposed samples for 15 min at 800 °C in air (31% relative humidity at 23 °C).

Figure 12.55 illustrates the evolution of the CMC mechanical properties after each degradation step. A decrease in ultimate tensile strength and work of fracture happens after the water exposure followed by a more pronounced embrittlement after the reheat.

The corresponding fracture surfaces from the water exposed and reheat samples are shown as SEM images in Fig. 12.56. The amount of fiber pullout in a fracture surface.

is an indication of the interfacial shear strength. The test results show a significantly higher pull-out length for the water exposed sample compared to the reheat one, which together with the mechanical results indicate that water degrades the fiber matrix interphase, which affect the samples after reheating. The main challenge is to understand how degradation affects the internal structure of the CMC and how these changes at the nano scale affect the mechanical properties. Micromechanical tests such as fiber push-out, micropillar compression or microcantilever beam fracture have been used to quantify the properties of the fiber-matrix interface. The push-out

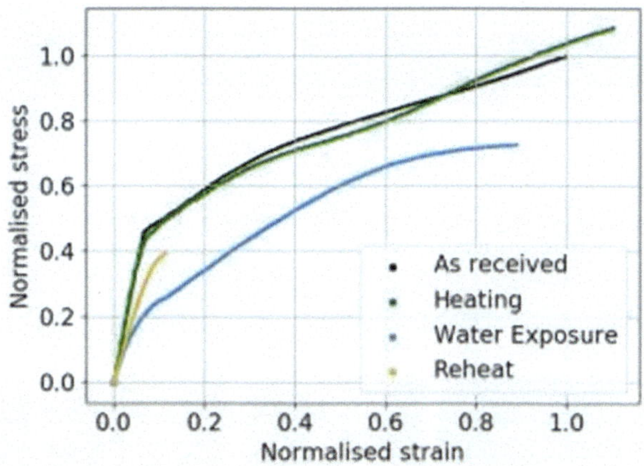

Fig. 12.55 Representative normalized stress–strain curve for the as-received material, after heating at intermediate temperatures (heated) after low temperature water exposure and after reheating at intermediate temperature. O. G. Diaz, K. Marquardt, S. Harris, L. Gale, L. Vandeperre, E. Saiz and F. Giuliani, J. Eur. Cer. Soc., 40, 3863 (2020). Open access

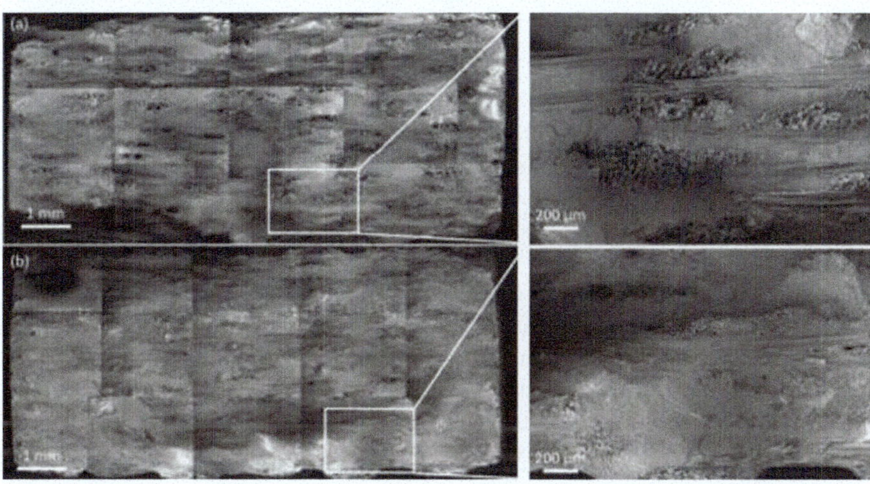

Fig. 12.56 Macro image of the fracture surfaces for: **a** Water Exposure and **b** Reheat samples. O. G. Diaz, K. Marquardt, S. Harris, L. Gale, L. Vandeperre, E. Saiz and F. Giuliani, J. Eur. Cer. Soc., 40, 3863 (2020). Open access

test has been chosen in this work to measure the Interfacial Shear Strength (ISS) and fiber friction during sliding. For the water exposed samples, fibers from the fractured surface of previously exposed and tested samples were used and compared with the as received and non-coated Hi Nicalon fibers.

The microstructure of the BN interphase did not change after the first heating step, showing a similar morphology to the one observed for the as received state as seen

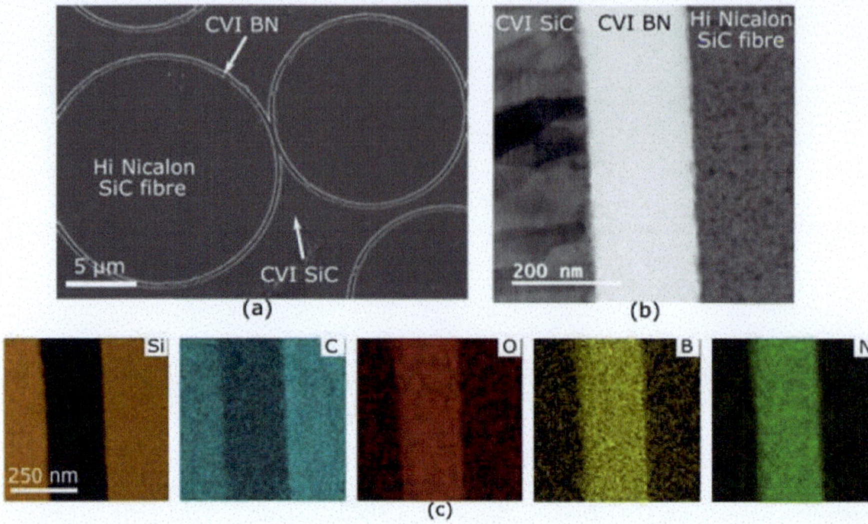

Fig. 12.57 Characterization of the effect of intermediate temperature on the degradation of BN interphases **a** SEM image of the overview, **b** Bright Field (BF) STEM image of the "as-heated" of the BN interphase and **c** EDS mapping of the STEM image. O. G. Diaz, K. Marquardt, S. Harris, L. Gale, L. Vandeperre, E. Saiz and F. Giuliani, J. Eur. Cer. Soc., 40, 3863 (2020). Open access

in Fig. 12.57 and compared with BN coated interface layer after water exposure in Fig. 12.58. Water exposure has a significant effect on the BN layer showing large sections of gaps (Fig. 12.58a) between the fiber and matrix. Microstructural changes are seen by STEM-EDS in the remaining interphase (Figs. 12.58b, c).

The reheating experiment after water exposure is conducted at 800 °C. The aim is to ascertain the degradation effect of intermediate temperatures after the interphase has been partially removed in the humid environment. The interfaces remain partially empty and there is further degradation as illustrated in Fig. 12.59.

The average interfacial shear stress applied to the interphase during in-situ push out tests in the SEM is calculated in as-received and degraded samples as

$$\tau = \frac{F}{2\pi rt} \tag{12.35}$$

where F is the force applied by the indenter, r the fiber radius and t the thickness of the sample. Further the tensile stress σ^{UTS} is calculated based on the position of the wedge (L.2) and the fiber radius (measured for each fiber tested) assuming a circular long beam by

$$\sigma^{UTS} = \frac{F_{Max}L}{\pi r^3} \tag{12.36}$$

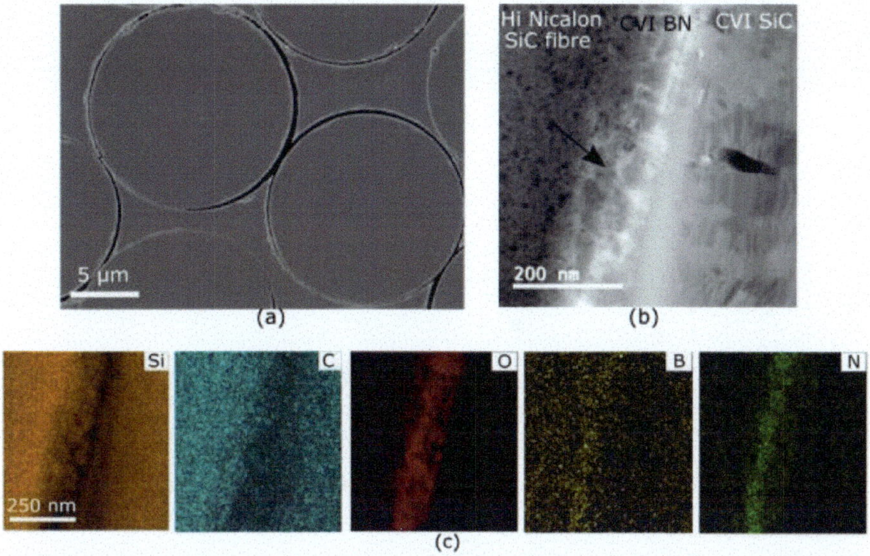

Fig. 12.58 Characterization of the effect of low temperature water exposure in the degradation of BN interphases: **a** SEM image of the overview showing how the interphase is partially leached out, **b** and **c** BF STEM and EDS analysis of the remaining interphase. O. G. Diaz, K. Marquardt, S. Harris, L. Gale, L. Vandeperre, E. Saiz and F. Giuliani, J. Eur. Cer. Soc., 40, 3863 (2020). Open access

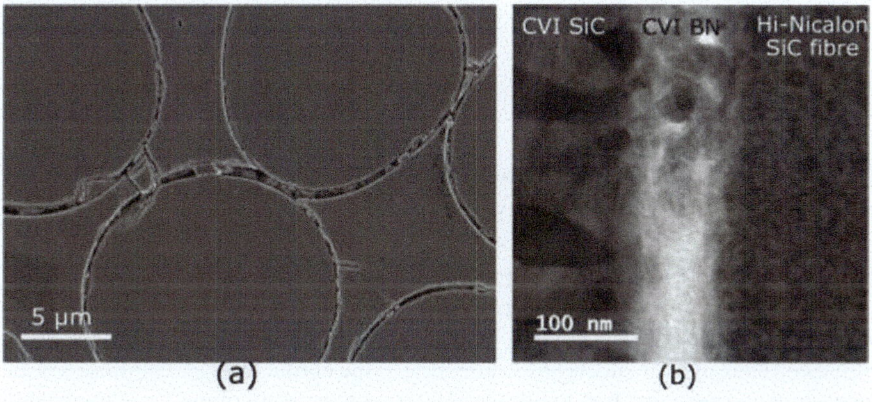

Fig. 12.59 **a** SE SEM and **b** BF STEM images after the reheat effect showing the alterations happening in the previously empty interfaces. O. G. Diaz, K. Marquardt, S. Harris, L. Gale, L. Vandeperre, E. Saiz and F. Giuliani, J. Eur. Cer. Soc., 40, 3863 (2020). Open access

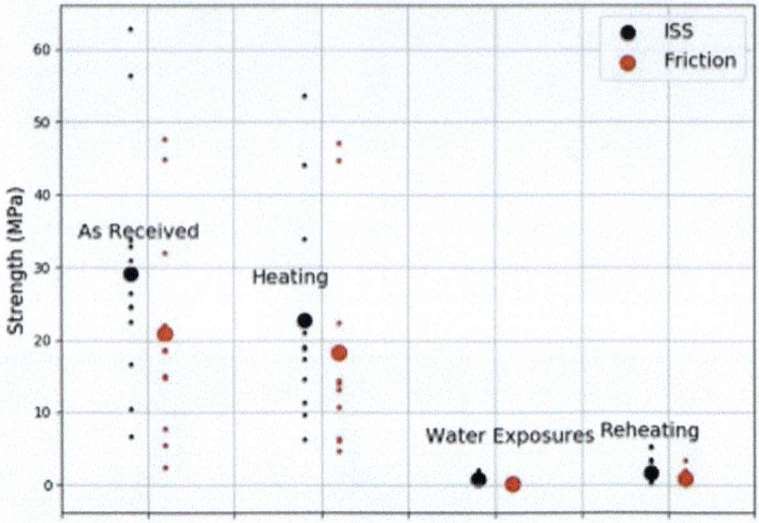

Fig. 12.60 Comparison of the Interfacial Shear Strength (ISS) and friction results found for the different steps of degradation: As received, Heating, Water Exposures and Reheating. The small dots represent individual tests and the large ones the average. O. G. Diaz, K. Marquardt, S. Harris, L. Gale, L. Vandeperre, E. Saiz and F. Giuliani, J. Eur. Cer. Soc., 40, 3863 (2020). Open access

P_{max} is the peak load of the force–displacement curve. The change in defect size, δ induced during reheating can be estimated based on linear elastic fracture mechanics:

$$K_{Ic} = \sigma(\pi d)^{1/2} \tag{12.37}$$

$$K_{IC}^{wat\ exp} = K_{Ic}^{reheat} \rightarrow \frac{\delta_{reheat}}{\delta_{wat\ exp}} = \left(\frac{\sigma_{wat\ exp}^{UTS}}{\sigma_{reheat}^{UTS}}\right) \tag{12.38}$$

where K_{Ic} and σ^{UTS} are the fracture toughness and the tensile strength

In summary, the mechanical properties, e.g., tensile strength and work of fracture of the SiC/BN/SiC CMCs are significantly affected by the low temperature water exposure. The material undergoes a reduction of the elastic limit and ultimate tensile strength. A subsequent short exposure at intermediate temperatures, say for 15 min at 800 °C, further affects the mechanical properties. It induces embrittlement with a massive reduction of the work of fracture and ultimate tensile strength of the fibers. Thus, as a result, subsequent intermediate temperature exposures (e.g. reheats in air) can trigger fiber degradation and significant drop in the mechanical response of the CMC, as illustrated in Fig. 12.55. More explicit information on water exposure SiC/BN/SiC CMCs and reheating is illustrated in Fig. 12.60 indicating the strength variation.

References

D.H. Alsem, O.N. Pierron, E.A. Stach, C.L. Muhlstein, R.O. Ritchie, Adv. Eng. Mater. **9**, 15 (2007)

P.A. Bartolotta, Enabling Propulsion Materials (EPM) Structural Component Successfully Tested Under Pseudo-Operating Conditions, Document ID 20050179367, Publication: Research and Technology (1996)

C.I. Contescu, Oak Ridge National Laboratory, ORNL/TM-2013/319 (2013)

X. Dai, F. Shen, Jia. Ji, S.L Wang, M. Xu, Mater. Sci. DeStech Trans. Eng. Techno. Rtes. (2018)

F.W. DelRio, R.F. Cook, B.L. Boyce, Appl. Phys. Rev. **2**, 021303 (2015)

O.G. Diaz, K. Marquardt, S. Harris, L. Gale, L. Vandeperre, E. Saiz, F. Giuliani, J. Eur. Cer. Soc. **40**, 3863 (2020)

J. Dong, W. Bai, G. Huang, L. Han, G. Zhang, Earth Environ. Sci. **571**, 012147 (2020)

T.-H. Fang, W.-J. Chang, C.-M. Lin, Microelect. Eng. **77**, 389 (2005)

X. Han, K. Zheng, Y.F. Zhang, X. Zhang, Z. Zhang, Z.L. Wang, Adv. Mater. **19**, 2112 (2007)

X.X. Li, T. Ono, Y.L. Wang, M. Esashi, Appl. Phys. Lett. **83**, 3081 (2003)

F. Lomello, G. Bonnefont, Y. Leconte, N. Herlin-Boime, G. Fantozzi, J. Eur. Ceram. Soc., **32**, 633 (2012)

W. Owens, D. Merkel, F. Sansoz, D. Fletcher, J. Am. Ceram. Soc. **98**, 4003 (2015)

A. Patel, E. Sato, T. Takagi, N. Shichijo, J. Eur. Cer. Soc. **42**, 4121 (2022)

J. Pelleg, Mechanical Properties of Nanomaterials, Springer Nature (2021)

K. Schulmeister, W. Mader, J. Non-Cryst, Solids **320**, 143 (2003)

M. Shankar, Mater. Sci. Eng. A **412**, 165 (2005)

L. Zhang, I. Zarudi, Wear **225**, 669 (1999)

Index

A
Activation energy
for electron traps in GaAs, 87
for releasing from oxygen bonding, 3, 122
in Si NW, 42, 163, 201

B
Bend test
in GaAs, 139, 184
in Si, 71, 121, 126, 154
in SiC, 185
Brittle, 149, 164
in Si, 179, 198

C
Compression, 14
in SiC, 52, 185, 210
Crack, 16, 26, 31, 51, 63, 71, 83, 91, 93, 100, 195, 198, 208, 209
Creep
in GaAs, 182
in SiC, 105, 107, 178, 194

D
Defect, 10, 26–28, 43, 63, 66, 79, 81, 85, 87, 96, 100, 134–141, 149, 152, 154, 209, 214
Deformation, 25, 44, 158, 166, 179, 201, 202
Ductility, 15
in SiC_1/SiC, 107
Dynamic, 43

E
Effect of moisture, 6, 31, 47, 67, 76, 93, 175, 179, 190, 209
Effect of oxygen, 2, 17, 25, 50, 71, 83, 105, 109, 121, 135, 154, 158, 163, 167, 177, 185, 192, 198
Environment, 1, 42, 50, 63, 71, 83, 105, 109, 121, 133, 149, 157, 163, 177, 185, 198

F
Fatigue, 6, 34, 109, 152, 179, 185, 200
Fracture, 210, 214
Fracture toughness, 10, 29, 30, 63, 71, 73, 91, 115, 149, 150, 152, 154, 207, 208, 214

G
Grain size, 6, 10, 17, 29, 31, 54, 92, 205
Growth, 107, 138, 152

H
Hall-Petch, 205
Hardness, 17, 36, 63, 69, 72, 90, 123, 136, 151, 154, 182, 205

I
Impact, 157
Indentation, 3, 10, 25, 36, 90, 122, 151, 182
Interstitial, 43–45, 49, 52, 122, 138, 154

J. Pelleg, *The Influence of Environment on Mechanical Characteristics of Key Semiconductor Components*, https://doi.org/10.1007/978-3-031-34530-2

N
Nano materials, 157, 163, 177, 185, 198
Nanowires, 102, 137, 138, 142, 163, 168, 199, 204
Notch effect, 152

T
Twins, 133, 134, 136, 139, 140

The manufacturer's authorised representative in the EU is Springer Nature Customer Service Centre GmbH, Europaplatz 3, 69115 Heidelberg, Germany. If you have any concerns regarding our products, please contact ProductSafety@springernature.com

Printed and bound by CPI Group (UK) Ltd, Croydon, CR0 4YY
31/07/2026
02182891-0001